Everyday Mathematics®

The University of Chicago School Mathematics Project

Student Math Journal
Volume 1

Grade 3

Wright Group

The University of Chicago School Mathematics Project (UCSMP)

Max Bell, Director, UCSMP Elementary Materials Component; Director, *Everyday Mathematics* First Edition
James McBride, Director, *Everyday Mathematics* Second Edition
Andy Isaacs, Director, *Everyday Mathematics* Third Edition
Amy Dillard, Associate Director, *Everyday Mathematics* Third Edition

Authors

Max Bell
Jean Bell
John Bretzlauf
Mary Ellen Dairyko*

Amy Dillard
Robert Hartfield
Andy Isaacs
James McBride

Kathleen Pitvorec
Peter Saecker

*Third Edition only

Technical Art
Diana Barrie

Teachers in Residence
Lisa Bernstein, Carole Skalinder

Editorial Assistant
Jamie Montague Callister

Contributors
Carol Arkin, Robert Balfanz, Sharlean Brooks, James Flanders, David Garcia, Rita Gronbach, Deborah Arron Leslie, Curtis Lieneck, Diana Marino, Mary Moley, William D. Pattison, William Salvato, Jean Marie Sweigart, Leeann Wille

Photo Credits
©Tim Flach/Getty Images, cover; ©Fotosearch, p. vii; Getty Images, cover, *bottom left;* ©Brian Hagiwara/Brand X Pictures/Getty Images, p. iii.

www.WrightGroup.com

Copyright © 2007 by Wright Group/McGraw-Hill.

All rights reserved. Except as permitted under the United States Copyright Act, no part of this publication may be reproduced or distributed in any form or by any means, or stored in a database or retrieval system, without the prior written permission from the publisher, unless otherwise indicated.

Printed in the United States of America.

Send all inquiries to:
Wright Group/McGraw-Hill
P.O. Box 812960
Chicago, IL 60681

ISBN 0-07-604567-6

13 CPC 13 12 11 10 09

Contents

UNIT 1 Routines, Review, and Assessment

Number Sequences 1
A Numbers Hunt 2
Number-Grid Puzzles 3
Looking up Information 4
Using Mathematical Tools 5
Displaying Data 6
Math Boxes 1✦5 8
Name-Collection Boxes 9
Math Boxes 1✦6 10
Can You Be Sure? 11
Math Boxes 1✦7 12
Finding Differences 13
Math Boxes 1✦8 14
Using a Calculator 15
Math Boxes 1✦9 16
Using Coins 17
Math Boxes 1✦10 19
A Shopping Trip 20
Coin Collections 21
Math Boxes 1✦11 22
Frames and Arrows 23
Patterns 24
Math Boxes 1✦12 25
Finding Elapsed Times 26
Sunrise and Sunset Record 27
Math Boxes 1✦13 28
Math Boxes 1✦14 29

Contents **iii**

UNIT 2 Adding and Subtracting Whole Numbers

Fact Families and Number Families 30
Math Boxes 2•1 31
Using Basic Facts to Solve Fact Extensions 32
Math Boxes 2•2 33
"What's My Rule?" 34
Math Boxes 2•3 35
Number Stories: Animal Clutches 36
Math Boxes 2•4 38
Number Stories: Change-to-More
    and Change-to-Less 39
Math Boxes 2•5 41
Temperature Differences 42
National High/Low Temperatures Project 43
Math Boxes 2•6 44
Addition Methods 45
Math Boxes 2•7 46
Subtraction Methods 47
Name-Collection Boxes 48
Math Boxes 2•8 49
Subtraction Methods 50
Number Stories with Several Addends 51
Math Boxes 2•9 53
Math Boxes 2•10 54

UNIT 3 Linear Measures and Area

Estimating and Measuring Lengths 55
Addition and Subtraction Practice 56
Math Boxes 3•1 57
Measuring Line Segments 58
Math Boxes 3•2 59
Measures Hunt 60
Estimating Lengths 61

Math Boxes 3◆3 62
Perimeters of Polygons 63
Body Measures 64
Math Boxes 3◆4 65
Math Boxes 3◆5 66
Geoboard Perimeters 67
Tiling with Pattern Blocks 68
Straw Triangles 70
Math Boxes 3◆6 71
Areas of Rectangles 72
Math Boxes 3◆7 73
More Areas of Rectangles 74
Math Boxes 3◆8 75
Diameters and Circumferences 76
Math Boxes 3◆9 77
Math Boxes 3◆10 78

UNIT 4 Multiplication and Division

Solving Multiplication Number Stories 79
Math Boxes 4◆1 80
More Multiplication Number Stories 81
Measuring Perimeter 82
Math Boxes 4◆2 83
Division Practice 84
Math Boxes 4◆3 85
Solving Multiplication
    and Division Number Stories 86
Math Boxes 4◆4 87
Subtraction Strategies 88
Math Boxes 4◆5 89
Math Boxes 4◆6 90
Math Boxes 4◆7 91
Exploration A: How Many Dots? 92
Exploration B: Setting Up Chairs 93
Math Boxes 4◆8 94
Estimating Distances 95
A Pretend Trip 96

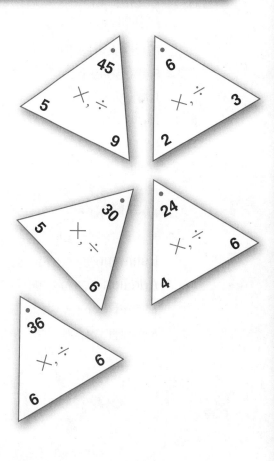

Contents **v**

Math Boxes 4•9 97
Coin-Toss Experiment 98
Measuring Line Segments 99
Math Boxes 4•10 100
Math Boxes 4•11 101

UNIT 5   Place Value in Whole Numbers and Decimals

Place-Value Review 102
Math Boxes 5•1 103
Math Boxes 5•2 104
Math Boxes 5•3 105
Working with Populations 106
Math Boxes 5•4 107
How Old Am I? 108
Math Boxes 5•5 109
Finding the Value of Base-10 Blocks 110
Squares, Rectangles, and Triangles 111
Pattern Block Perimeters 112
Math Boxes 5•6 113
Place Value in Decimals 114
Math Boxes 5•7 116
Exploring Decimals 117
Math Boxes 5•8 118
Decimals for Metric Measurements 119
Math Boxes 5•9 120
How Wet? How Dry? 121
Math Boxes 5•10 122
More Decimals 123
Math Boxes 5•11 124
Length-of-Day 125
Math Boxes 5•12 126
Math Boxes 5•13 127

UNIT 6   Geometry

Line Segments, Rays, and Lines 128

Math Boxes 6•1 129
Geometry Hunt 130
Math Boxes 6•2 131
Turns 132
Math Boxes 6•3 133
Exploring Triangles 134
Math Boxes 6•4 135
Exploring Quadrangles 136
Math Boxes 6•5 137
Exploring Polygons 138
Math Boxes 6•6 140
Drawing Angles 141
Math Boxes 6•7 142
Marking Angle Measures 143
Measuring Angles 144
Math Boxes 6•8 145
Symmetric Shapes 146
Math Boxes 6•9 147
Base-10 Block Decimal Designs 148
10 × 10 Grids 149
Math Boxes 6•10 150
Symmetry 151
Math Boxes 6•11 152
Pattern-Block Prisms 153
Math Boxes 6•12 154
Math Boxes 6•13 155

Activity Sheets

Multiplication/Division
    Fact Triangles 1 .... **Activity Sheet 1**
Multiplication/Division
    Fact Triangles 2 .... **Activity Sheet 2**
×, ÷ Fact Triangles 3 .... **Activity Sheet 3**
×, ÷ Fact Triangles 4 .... **Activity Sheet 4**

Contents **vii**

LESSON 1·1   Number Sequences

Complete the number sequences.

1. 428, 429, _____, 431, _____, _____, _____, _____, 436, _____, ...

   Unit

2. 918, 919, _____, _____, 922, _____, _____, _____, 926, ...

3. _____; 1,416; _____; _____; 1,419; _____; _____; ...

4. _____, 311, _____, _____, 341, _____, _____, ...

5. _____; 4,326; _____; _____; 4,356; _____; ...

Try This

6. 7,628; _____; 7,828; _____; _____; 8,128

one **1**

Date _____ Time _____

LESSON 1·1 A Numbers Hunt

Look for numbers in your classroom. Write the numbers in the table. Look for numbers that you cannot see but you can find by counting or measuring. Record these numbers, too.

Number	Unit (if there is one)	What does the number tell you?	How did you find the number? (count, measure, another way?)
16	Crayons	Tells how many crayons are in a box	Number is on the box
30	Inches	Height of my desk	Measured my desk

Date _____ Time _____

LESSON 1·2 Number-Grid Puzzles

1. Follow your teacher's directions.

541			544						550
551		553			556			559	
	562			565					570
			574				577		
581				585				588	
		593						599	
	602				606				
			614						620

Fill in the missing numbers.

2.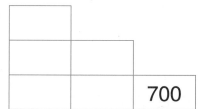

3.

4.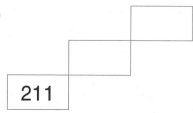

5.

6.

7.

Do your own.

8.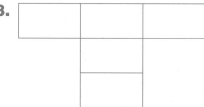

9.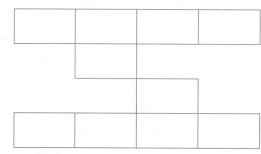

three **3**

Date _____ Time _____

LESSON 1·3   Looking up Information

1. Turn to page 246 in your *Student Reference Book*.

   How many yards are there in 1 mile? _____ yards

Work with a partner. Use your *Student Reference Book* for Questions 3–6.

2. Write your partner's first name. _____

   Write your partner's last name. _____

3. Look up the word **circumference** in the Glossary. Copy the definition.

   _____

4. Read the essay "Tally Charts."

   a. Then solve the Check Your Understanding problems.

      Problem 1: _____

      Problem 2: _____

   b. Check your answers in the Answer Key.

   c. Describe what you did to find the essay.

      _____

      _____

5. Find the Measurement section. Which of the following units
   of length is about the same length as a person's height? _____

   a. yard   b. thumb   c. fathom   d. cubit   e. hand   f. foot

   On which page did you find the answer? _____

6. Look up the rules of the game *Less Than You!* Play the game with
   your partner.

4   four

Date _____ Time _____

LESSON 1·4 Using Mathematical Tools

In Problems 1 and 2, record the time shown on the clocks. In Problem 3, draw the minute hand and the hour hand to show the time.

1.

2.

3.

_____   _____   6:10

Use your ruler.

4. Measure the line segment.   about _____ inches

_____

5. Draw a line segment 10 centimeters long.

Use your calculator to do these problems.

6. 23,573 + 859 + 6,051 = _____

7. 20,748 − 8,967 = _____

8. 466 × 38 = _____

9. 1,978 ÷ 23 = _____

Use your Pattern-Block Template to draw the following shapes:

10. a rhombus

11. a hexagon

12. a trapezoid

Try This

13. Which of the shapes in Problems 10–12 are quadrangles?

_____

five   5

Date _____ Time _____

LESSON 1·5 Displaying Data

1. How many first names are there? _____ last names? _____

2. Make a tally chart for either first names or last names.

_____ Names	
Number of Letters	Number of Children
2	
3	
4	
5	
6	
7	
8	
9 or more	

3. How many letters does the longest name have? _____ letters
   The number of letters in the longest name is called the **maximum.**

4. How many letters does the shortest name have? _____ letters
   The number of letters in the shortest name is called the **minimum.**

5. What is the **range** of the numbers of letters? _____ letters
   The range is the difference between the largest and smallest numbers.

6. What is the **mode** of the set of data? _____ letters
   The number that occurs most often is called the mode.

7. What is the **median** of the set of data? _____ letters
   (*Hint:* Look in your *Student Reference Book*.)

LESSON 1·5 Displaying Data *continued*

8. Make a bar graph for your set of data.

Title: _____

LESSON 1·5 Math Boxes

1. Write the numbers that are 10 less and 10 more.

   _____ 38 _____

   _____ 245 _____

   _____ 367 _____

   _____ 1,587 _____

2. Put these numbers in order from smallest to largest.

   306 _____ (smallest)

   296 _____

   496 _____

   936 _____ (largest)

3. Write 5 names in the 25-box.

4. Draw hands on the clock to show 6:45.

5. Lara brought 14 candies to school. She gave away 7 during recess. How many candies does she have now?

   _____ candies

6. Add.

   Unit

   0 + 7 = _____

   5 + 1 = _____

   3 + 3 = _____

   _____ = 4 + 7

   _____ = 9 + 6

8 eight

Date   Time

LESSON 1·6   Name-Collection Boxes

Work with a partner.

1. Write at least 10 names in the 20-box.

2. Three names do not belong in this box. Cross them out. Then write the name of the box on the tag.

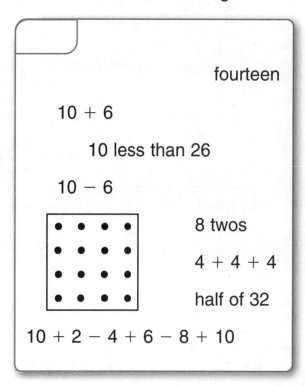

fourteen

$10 + 6$

10 less than 26

$10 - 6$

8 twos

$4 + 4 + 4$

half of 32

$10 + 2 - 4 + 6 - 8 + 10$

Do these on your own.

3. Write at least 10 names in the 24-box.

4. Make up your own box. Write at least 10 names.

nine 9

Date _____ Time _____

LESSON 1·6 Math Boxes

1. Measure the line segment to the nearest inch.

   _____ inches

   to the nearest centimeter

   _____ centimeters

   _____

2. Fill in the missing numbers.

	352

3. Put these numbers in order from smallest to largest.

   2,764   _____

   8,596   _____

   2,199   _____

   5,096   _____

4. Count back by 2s.

   104, _____, _____, _____,

   96, _____, _____, _____,

   _____, _____, _____, _____,

   _____, _____, _____, _____

5. 6,347

   What value does the 6 have? <u>6,000</u>

   What value does the 3 have? _____

   What value does the 4 have? _____

   What value does the 7 have? _____

6. Marque had $6. His mother gave him $8. How much money does Marque have now?

   $_____

10 ten

Date   Time

LESSON 1·7   Can You Be Sure?

1. Make a list of things you are *sure* will happen.

   _____
   _____
   _____
   _____
   _____
   _____

2. Make a list of things you are sure will *not* happen.

   _____
   _____
   _____
   _____
   _____

3. Make a list of things you think *might* happen but you are not sure about.

   _____
   _____
   _____
   _____
   _____
   _____

eleven   **11**

LESSON 1·7 Math Boxes

1. Write the numbers.

   10 less   10 more

   _____ 100 _____

   _____ 200 _____

   _____ 300 _____

   _____ 1,000 _____

2. Put these numbers in order from smallest to largest.

   4,306 _____

   4,296 _____

   3,496 _____

   2,936 _____

3. Write at least 5 names in the 75-box.

4. What time does the clock show?

   _____

5. Allison swam 16 laps in the pool. Carmen swam 9. How many more laps did Allison swim than Carmen? Fill in the circle next to the best answer.

   ○ A. 25 laps

   ○ B. 23 laps

   ○ C. 16 laps

   ○ D. 7 laps

6. Add.

   3 + 6 = _____

   _____ = 5 + 7

   8 + 6 = _____

   9 + 9 = _____

   6 + 4 = _____

   Unit

12 twelve

Date _____ Time _____

LESSON 1·8 Finding Differences

									0
1	2	3	4	5	6	7	8	9	10
11	12	13	14	15	16	17	18	19	20
21	22	23	24	25	26	27	28	29	30
31	32	33	34	35	36	37	38	39	40
41	42	43	44	45	46	47	48	49	50
51	52	53	54	55	56	57	58	59	60
61	62	63	64	65	66	67	68	69	70
71	72	73	74	75	76	77	78	79	80
81	82	83	84	85	86	87	88	89	90
91	92	93	94	95	96	97	98	99	100
101	102	103	104	105	106	107	108	109	110

Use the number grid to help you solve these problems.

1. Which is less, 83 or 73? _____ How much less? _____

2. Which is less, 13 or 34? _____ How much less? _____

3. Which is more, 90 or 55? _____ How much more? _____

4. Which is more, 44 or 52? _____ How much more? _____

Find the **difference** between each pair of numbers.

5. 71 and 92 _____   6. 26 and 46 _____

7. 30 and 62 _____   8. 48 and 84 _____

9. 43 and 60 _____   10. 88 and 110 _____

thirteen **13**

Date _____ Time _____

LESSON 1·8 Math Boxes

1. Measure the line segment

   to the nearest inch. _____ in.

   to the nearest centimeter.

   _____ cm

   _____

2. Fill in the missing numbers.

   632

   644

3. Put these numbers in order from largest to smallest.

   2,764 _____

   596 _____

   2,199 _____

   8,096 _____

4. Count by 2s.

   1,012; 1,014; _____;

   _____; _____; _____;

   _____; _____; _____;

   _____; _____; _____;

   _____; _____; _____

5. 1,942

   What value does the 4 have? _____

   What value does the 9 have? _____

   What value does the 1 have?

   _____

   What value does the 2 have? _____

6. Andre scored 7 points. Tina scored 5 points. How many points did they score altogether? Fill in the circle next to the best answer.

   ○ **A.** 2 points

   ○ **B.** 12 points

   ○ **C.** 17 points

   ○ **D.** 35 points

14 fourteen

Date _____ Time _____

LESSON 1·9 Using a Calculator

Math Message

Use your calculator.

1. Sharon read the first 115 pages of her book last week. She read the rest of the book this week. If she read 86 pages this week, how many pages long is her book?

    Answer: Her book is _____ pages long.

    Number model: _____

2. The paper clip was invented in 1868. The stapler was invented in 1900. How many years after the paper clip was the stapler invented?

    Answer: The stapler was invented _____ years later.

    Number model: _____

3. 28 + 64 + 39 + 42 = _____   4. 2,648 − 1,576 = _____

Calculator Practice

Use your calculator.

5. Begin at 25. Count up by 6s. Record your counts below.

    25 ___ ___ ___ ___ ___ ___ ___ ___

6. Begin at 90. Count back by 9s.

    90 ___ ___ ___ ___ ___ ___ ___ ___

Solve the calculator puzzles. Remember to add or subtract to find the "Change to" number.

7. | Enter | Change to | How? |
    |---|---|---|
    | 42 | 52 | _____ |
    | 61 | 41 | _____ |
    | 145 | 105 | _____ |

8. | Enter | Change to | How? |
    |---|---|---|
    | 178 | 208 | _____ |
    | 1,604 | 804 | _____ |
    | 722 | 3,722 | _____ |

fifteen **15**

LESSON 1·9 Math Boxes

1. How many children chose grapes? _____ children
   How many children chose oranges? _____ children

   Children's Fruit Choices

2. Count back by 3s.

   42, _____, _____, 33,

   _____, _____, _____, _____,

   _____, _____, _____, _____,

   _____, _____, _____, _____

3. Use + or − to make each number sentence true.

   8 = 13 _____ 5

   15 = 7 _____ 8

   17 _____ 9 = 8

4. Circle the letter next to the coins that do **not** show $0.89.

   A Q Q Q D P P P P

   B Q Q Q D D P P P P P

   C Q Q D D D P P P P P P P

   D Q Q D D D N P P P

5. What is today's date?

   _____

   What will be the date in 6 days?

   _____

   What will be the date in 1 week?

   _____

6. Fill in the blanks.   Unit

   8 + _____ = 15

   7 + _____ = 15

   _____ − 8 = 7

   15 − _____ = 8

Date _____ Time _____

Lesson 1·10 Using Coins

Math Message

1. You buy a carton of juice for 89 cents. Show two ways to pay for it with exact change. Draw Ⓟ to show pennies, Ⓝ to show nickels, Ⓓ to show dimes, and Ⓠ to show quarters.

   a. _____ b. _____

   _____

Write each of the following amounts in dollars-and-cents notation. The first one is done for you.

2. three dimes and one nickel $0.35

3. five dimes and seven pennies _____

4. fourteen dimes _____

5. two quarters and four pennies _____

6. three dollars and one nickel and three pennies _____

7. seven dollars and eight dimes _____

Write =, <, or >.

8. $0.68 _____ ⓆⓆⓆ

9. ⒹⒹⓃⓃⓅⓅ _____ ⓆⓃⓅ

10. $1.18 _____ $1.81

11. three quarters _____ three dimes

12. ten dimes _____ one dollar

13. $0.67 _____ seven dimes

Remember

= means *is equal to*

< means *is less than*

> means *is greater than*

seventeen **17**

LESSON 1·10

Using Coins continued

14. Circle the digit that represents dimes.

$17.6 3

15. Circle the digit that represents pennies.

$18.34

16. Circle the digit that represents dimes.

3 5 ¢

17. Jean wants to buy a carton of milk for 35¢.
How much change will she get from 2 quarters? _____

Use Ⓠ, Ⓓ, Ⓝ, and Ⓟ to show her change in two ways.

_____

Try This

Use the Drinks Vending Machine Poster on *Student Reference Book,* page 212.

18. Marcy wants to get a strawberry yogurt drink and a chocolate milk from the vending machine. She has only dollar bills.

a. If the Exact Change light is on, can she buy what she wants? _____

b. If the Exact Change light is off, how many dollar bills will she put in the machine? _____

How much change will she get? _____

18 eighteen

Date   Time

LESSON 1·10

Math Boxes

1. Write the number that is 10 more.

   42 _____

   160 _____

   901 _____

   59 _____

   120 _____

2. Ages of 9 teachers: 30, 24, 49, 50, 38, 44, 40, 35, 51

   median = _____

   maximum = _____

3. Write at least 5 names in the 1-box.

4. Describe 2 events that are impossible.

5. Put these numbers in order from smallest to largest.

   7,912 _____

   7,192 _____

   9,271 _____

   9,172 _____

6. Fill in the missing numbers.

   _____ = 3 + 5

   _____ = 8 − 5

   _____ = 5 + 3

   _____ = 8 − 3

   Unit

nineteen **19**

Date _____ Time _____

LESSON 1·11 A Shopping Trip

Use the Stationery Store Poster on *Student Reference Book,* page 214.

1. List the items you are buying in the space below. You must buy at least 3 items. You can buy 2 of the same item but list it twice.

Item	Sale Price
_____	_____
_____	_____
_____	_____

2. Estimate how many dollar bills you will need to give the shopkeeper to pay for your items.   _____ dollar bills.

3. Give the shopkeeper the dollar bills.

4. The shopkeeper calculates the total cost using a calculator.

    You owe $_____.

5. The shopkeeper calculates the change you should be getting. $_____

6. Use Ⓟ, Ⓝ, Ⓓ, Ⓠ, and $1 to show the change you got from the shopkeeper. _____

Try This

7. Henry buys one pack of batteries and a box of crayons. How much money does he save buying them on sale instead of paying the regular price?

	Regular Price	Sale Price		Difference
batteries	$_____ . _____	$_____ . _____	Regular total	$_____ . _____
crayons	$_____ . _____	$_____ . _____	Sale total	$_____ . _____
Total Cost	$_____ . _____	$_____ . _____	Amount Saved	$_____ . _____

20 twenty

LESSON 1·11 Coin Collections

Get your coin collection or grab a handful of coins from the classroom collection. Complete the problems below.

1. Count each kind of coin. Give a total value for each type of coin.

   _____ Ⓟ = $_____._____

   _____ Ⓝ = $_____._____

   _____ Ⓓ = $_____._____

   _____ Ⓠ = $_____._____

2. What is the total value of all the coins? You may use a calculator.

   Total value = $_____._____

3. In the space below, draw a picture of your total. Use as few $1, Ⓠ, Ⓓ, Ⓝ, and Ⓟ as possible.

Try This

4. Explain how you would enter your total amount on the calculator.

   _____

   _____

5. How much money would you need to go up to the next dollar amount? (*Hint:* A dollar amount is $1.00, $2.00, $3.00, and so on.)

   _____

   _____

6. Explain how you would go up to the next dollar amount without clearing your calculator.

   _____

   _____

twenty-one **21**

Date _____ Time _____

LESSON 1·11 Math Boxes

1. How many children chose apples? _____ children

   How many children chose pears? _____ children

   Children's Fruit Choices

   (dot plot: apples 4, grapes 5, oranges 3, pears 8)

   Fruit Choices

   SRB 77 78

2. Count by 10s.

   23, _____, _____, 53, _____,

   _____, _____, _____, _____,

   _____, _____, _____, _____

3. Use + or − to make each number sentence true.

   5 ____ 6 = 11

   6 ____ 5 = 1

   14 = 7 ____ 7

4. Use Q, D, N, and P. Show $1.48 in two ways.

5. What is today's date?

   _____

   What will be the date in 11 days?

   _____

   What will be the date in 2 weeks?

   _____

6. Complete the fact triangle.

   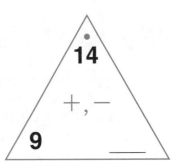

22 twenty-two

Lesson 1·12 Frames and Arrows

Math Message

Find the pattern. Fill in the missing numbers.

1. 37, 40, 43, ____, ____, ____
2. 27, 25, ____, 21, ____, ____
3. ____, 11, 15, ____, 23, ____
4. ____, ____, 36, 33, ____, 27

Frames and Arrows

5. Rule: +5¢
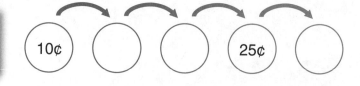

6. Rule: +10
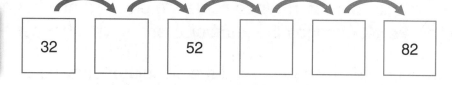

7. Rule:
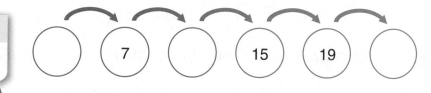

8. Rule: Double
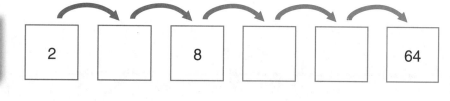

9. Make up one of your own.

 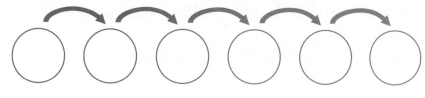

twenty-three 23

LESSON 1·12 Patterns

Complete the number-grid puzzles.

1. 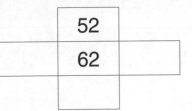   2. 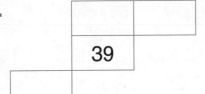   3.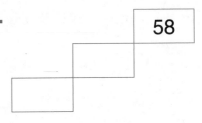

4. Draw dots to show what comes next.

5. Janie owns a magic calculator. When someone enters a number and then presses the ⊜ key, it changes the number. Here is what happened:

   ♦ Tom entered 15. He pressed ⊜ and the calculator showed 5.

   ♦ Mary entered 12. She pressed ⊜ and the calculator showed 2.

   ♦ Regina entered 27. She pressed ⊜ and the calculator showed 17.

6. What do you think the calculator will show if Janie enters 109 and ⊜? _____

7. Explain how you know. _____

Try This

8. The numbers below have a pattern. Fill in the missing numbers. Be careful: The same thing does not always happen each time.

   4, 14, 24, 22, 32, 42, 40, 50, 60, 58, _____, _____, _____

9. Describe the pattern. _____

Date _____ Time _____

LESSON 1·12 Math Boxes

1. Write the number that is 100 more.

    237 _____

    614 _____

    994 _____

    2,462 _____

    3,965 _____

2. Median number of books read: ____ books

    Maximum number of books read: ____ books

    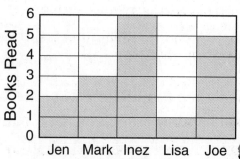

3. Write 5 names in the 0-box.

4. Describe 2 events that are almost sure to happen.

    _____
    _____
    _____
    _____

5. Which group is in order from largest to smallest? Fill in the circle next to the best answer.

    Ⓐ 4,039; 4,040; 4,409; 4,009

    Ⓑ 4,409; 4,040; 4,039; 4,009

    Ⓒ 4,009; 4,039; 4,040; 4,409

    Ⓓ 4,040; 4,039; 4,009; 4,409

6. Fill in the blanks.

    Unit

    4 + 2 = ____

    6 − ____ = 2

    ____ + 4 = 6

    6 − 2 = ____

twenty-five **25**

Date _____ Time _____

LESSON 1·13 Finding Elapsed Times

Use your toolkit clock to help you solve these problems.

1. Ava leaves to go swimming at 4:05 and returns at 5:25. How long has she been gone? _____

2. Deven rides his bike 37 miles. He rides from 10:15 A.M. until 3:50 P.M. How long does it take him to ride 37 miles? _____

3. LaToya leaves for school at the time shown on the first clock. She returns home at the time shown on the second clock. How long is LaToya away from home?

    _____

Try This

4. Gregory baked cookies for a class party. He baked several different kinds. He began baking at the time shown on the first clock and finished at the time shown on the second clock. How long did it take Gregory to bake the cookies?

    _____

Explain how you figured out the answer.

_____

_____

_____

_____

Date   Time

LESSON 1·13 Sunrise and Sunset Record

Date	Time of Sunrise	Time of Sunset	Length of Day	
			hr	min
			hr	min
			hr	min
			hr	min
			hr	min
			hr	min
			hr	min
			hr	min
			hr	min
			hr	min
			hr	min
			hr	min
			hr	min
			hr	min
			hr	min
			hr	min
			hr	min
			hr	min
			hr	min
			hr	min
			hr	min
			hr	min
			hr	min
			hr	min
			hr	min

Date _____ Time _____

LESSON 1·13 Math Boxes

1. How many children like lions? _____

   How many children like tigers? _____

Animal choice	Number of Children
Bears	////
Lions	＃＃ /
Crows	///
Tigers	＃＃ ＃＃

2. Count back by 4s.

   104, _____, _____, _____, 88

   _____, _____, _____, _____,

   _____, _____, _____, _____,

   _____, _____, _____

3. Use + or – to make each number sentence true.

   9 = 3 ____ 3 ____ 3

   12 = 4 ____ 4 ____ 4

   18 = 9 ____ 9

4. Draw the bills and coins to show $2.43 in two ways.

5.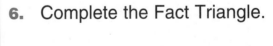

   If October 6 is on a Sunday, what are the dates for the *next* two Sundays?

   _____

   _____

6. Complete the Fact Triangle.

   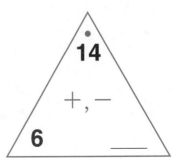

28 twenty-eight

Lesson 1·14 Math Boxes

1. Complete the fact family.

4 + 5 = ___

5 + ___ = 9

9 − 4 = ___

___ − 5 = ___

Unit

2. Ava scored 9 goals this season. Jamar scored 6 goals. How many goals did they score altogether?

_____ goals

3. Fill in the blanks.

7 − _____ = 7

13 + _____ = 13

9 + 1 = _____

8 + 1 = _____

Unit

4. Nico walks 6 blocks to school. Cyrus walks 4 blocks to school. How many blocks do they walk in all?

_____ blocks

5. Fill in the blanks.

_____ = 9 − 8

_____ = 7 − 6

1 = 5 − _____

1 = _____ − 7

Unit

6. Jeri brought 10 erasers to school. She gave 6 erasers to her friends. How many erasers does she have left?

_____ erasers

twenty-nine 29

Date _____ Time _____

LESSON 2·1 Fact Families and Number Families

Complete the Fact Triangles. Write the fact families.

1.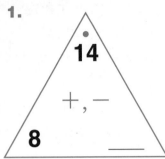

2.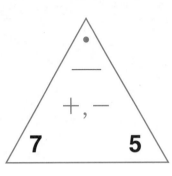

3.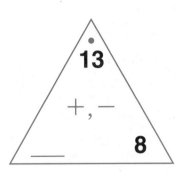

___ + ___ = ___   ___ = ___ + ___   ___ + ___ = ___

___ + ___ = ___   ___ = ___ + ___   ___ + ___ = ___

___ − ___ = ___   ___ = ___ − ___   ___ − ___ = ___

___ − ___ = ___   ___ = ___ − ___   ___ − ___ = ___

Complete the number triangles. Write the number families.

4.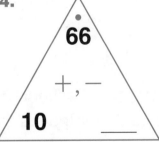

5.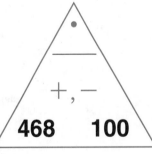

6.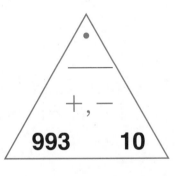

___ = ___ + ___   ___ = ___ + ___   ___ + ___ = ___

___ = ___ + ___   ___ = ___ + ___   ___ + ___ = ___

___ = ___ − ___   ___ = ___ − ___   ___ − ___ = ___

___ = ___ − ___   ___ = ___ − ___   ___ − ___ = ___

30 thirty

Date _____ Time _____

LESSON 2·1 Math Boxes

1. Put these numbers in order from smallest to largest.

1,532 _____

1,253 _____

1,325 _____

5,321 _____

2. Fill in the missing numbers.

		975

(grid extending downward below 975)

3. Write the numbers that are 10 less and 10 more than each given number.

         10 less   10 more

368 _____ _____

789 _____ _____

1,999 _____ _____

40,870 _____ _____

4. I spent $3.25 at the store. I gave the cashier a $5.00 bill. How much change should I have received?

_____

5. About what time is it? Fill in the circle next to the best answer.

Ⓐ 1:35

Ⓑ 7:08

Ⓒ 1:40

Ⓓ 2:35

6. Measure the line segment in inches.

_____ inches

_____

thirty-one **31**

Date _____ Time _____

LESSON 2·2 Using Basic Facts to Solve Fact Extensions

Fill in the unit box. Complete the fact extensions.

Unit: _____

1. _____ = 12 − 7
   _____ = 120 − 70
   _____ = 1,200 − 700

2. 8 + 3 = _____
   80 + 30 = _____
   800 + 300 = _____

3. _____ = 7 + 6
   _____ = 70 + 60
   _____ = 700 + 600

Complete the fact extensions.

4. _____ = 6 + 8
   _____ = 16 + 8
   _____ = 56 + 8

5. 14 − 9 = _____
   24 − 9 = _____
   54 − 9 = _____

6. _____ = 17 − 11
   _____ = 27 − 11
   _____ = 47 − 11

Use addition or subtraction to complete these problems on your calculator. You may also use a number grid, base-10 blocks, or counters.

7.
Enter	Change to	How?
33	40	_____
80	73	_____
80	23	_____

8.
Enter	Change to	How?
430	500	_____
700	640	_____
1,000	400	_____

9. Why is it important to know the basic addition and subtraction facts?

_____

_____

Date _____ Time _____

Lesson 2·2 Math Boxes

1. Complete the Fact Triangle. Write the fact family.

   ____ + ____ = ____

   ____ + ____ = ____

   ____ − ____ = ____

   ____ − ____ = ____

   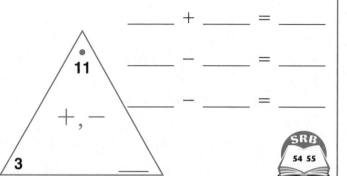

2. Choose the best answer.

   The school chorus has 28 second graders and 34 third graders. How many children are in the chorus?

   ⚪ 52 children   ⚪ 14 children

   ⚪ 12 children   ⚪ 62 children

3. Use your calculator. Write the answers in dollars and cents.

   $0.85 + 53¢ = _____

   $2.08 + $5.01 = _____

   64¢ + $1.73 = _____

   37¢ + 26¢ = _____

4. Find the rule. Fill in the empty frames.

   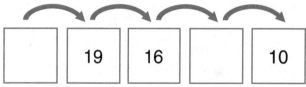

5. The mode for the number of books read is _____.

   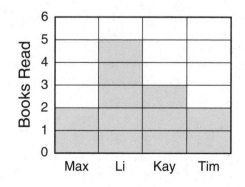

6. Write these units of measure in order from smallest to largest.

   mile   _____ ← smallest

   foot   _____

   yard   _____

   inch   _____ ← largest

thirty-three **33**

Date   Time

LESSON 2·3 "What's My Rule?"

Fill in the blanks.

1.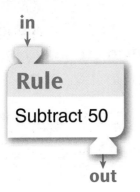

in	out
100	
50	
70	
150	
200	

2.

in	out
14	23
34	43
44	53
64	73
94	103

3.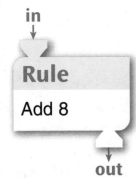

in	out
	13
	23
	43
	73
	93

4. Rule: Subtract 30

in	out
	30
	50
	100
	200
	0

5.

in	out
6	13
9	
5	
4	11
	18

6.

in	out
35	20
	45
20	
50	35
46	

34   thirty-four

| Date | Time |

LESSON 2·3 Math Boxes

1. Put these numbers in order from smallest to largest.

   32,764 _____

   8,596 _____

   32,199 _____

   85,096 _____

2. Fill in the missing numbers.

	1,073		

	1,104

3. Write the numbers that are 100 more and 100 less than each given number.

           100 more   100 less

   614 _____ _____

   994 _____ _____

   2,462 _____ _____

   3,965 _____ _____

4. You spent $7.88 at the store. You gave the cashier a $10 bill. How much change should you receive?

   _____

5. What time is it?

   _____

   What time will it be in 20 minutes?

   _____

   How many minutes until 5:15?

   _____

6. Measure the line segment to the nearest $\frac{1}{2}$ inch.

   _____ inches

   _____

thirty-five 35

LESSON 2·4 Number Stories: Animal Clutches

For each number story, write ? for the number you want to find. Write the numbers you know in the parts-and-total diagram. Solve the problem, and write a number model.

1. Two pythons laid clutches of eggs. One clutch had 36 eggs. The other had 23 eggs. How many eggs were there in all?

   Answer the question: _____
   (unit)

   Number model: _____

   Check: How do you know your answer makes sense?

   _____

   _____

Total	
Part	Part

2. A queen termite laid about 6,000 eggs on Monday and about 7,000 eggs on Tuesday. About how many eggs did she lay in all?

   Answer the question: _____
   (unit)

   Number model: _____

   Check: How do you know your answer makes sense?

   _____

   _____

Total	
Part	Part

3. Two clutches of Mississippi alligator eggs were found. Each clutch had 47 eggs. What was the total number of eggs found?

   Answer the question: _____
   (unit)

   Number model: _____

   Check: How do you know your answer makes sense?

   _____

Total	
Part	Part

Date   Time

LESSON 2·4 Number Stories: Animal Clutches continued

Try This

4. An alligator clutch had 60 eggs. Only 12 hatched. How many eggs did not hatch?

   Answer the question: _____ (unit)

   Number model: _____

   Check: How do you know your answer makes sense?

   _____

   _____

Total	
Part	Part

5. Scientists say a green turtle can lay about 1,800 eggs in a lifetime, but only about 400 eggs hatch overall. About how many eggs do not hatch?

   Answer the question: _____ (unit)

   Number model: _____

   Check: How do you know your answer makes sense?

   _____

   _____

Total	
Part	Part

thirty-seven **37**

Date _____ **Time** _____

LESSON 2·4 Math Boxes

1. Complete the Fact Triangle. Write the fact family.

____ + ____ = ____

____ + ____ = ____

____ − ____ = ____

____ − ____ = ____

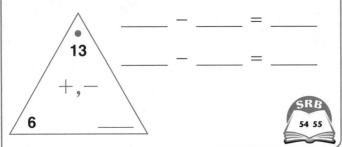

SRB 54 55

2. Jonah had $52. He bought a CD for $14. How much money does he have now?

_____

Number model:

_____

3. Use your calculator. Write the answers in dollars and cents.

73¢ + $2.65 = _____

$0.65 + 47¢ = _____

$1.06 + $5.21 = _____

46¢ + 35¢ = _____

4. Fill in the empty frames.

Rule
+100

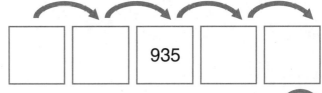

☐ ☐ 935 ☐ ☐

SRB 200 201

5. Weekly allowances: $15, $12, $5, $8

The maximum weekly allowance is _____.

The minimum weekly allowance is _____.

The range of weekly allowances is

 .

SRB 79

6. Write these metric units of measure in order from smallest to largest.

centimeter _____ ← smallest

kilometer _____

millimeter _____

meter _____ ← largest

SRB 141

38 thirty-eight

Date _____ Time _____

LESSON 2·5 Number Stories: Change-to-More and Change-to-Less

For each number story, write ? in the diagram for the number you want to find. Then write the numbers you know in the change diagram also. Next, solve the problem. Write the answer and a number model.

Unit: dollars

1. Ahmed had $22 in his bank account. For his birthday, his grandmother deposited $25 for him. How much money is in his bank account now?

   Answer the question: _____

   Number model: _____

   Check: How do you know your answer makes sense?

   _____

2. Omar had $53 in his piggy bank. He used $16 to take his sister to the movies and buy treats. How much money is left in his piggy bank?

   Answer the question: _____

   Number model: _____

   Check: How do you know your answer makes sense?

   _____

3. Cleo had $37 in her purse. Then Jillian returned $9 that she borrowed. How much money does Cleo have now?

   Answer the question: _____

   Number model: _____

   Check: How do you know your answer makes sense?

   _____

thirty-nine **39**

Lesson 2·5 Number Stories continued

4. Audrey had $61 in her bank account. She withdrew $48 to take on vacation. How much is left in her account?

   Answer the question: _____

   Number model: _____

   Check: How do you know your answer makes sense?

   _____

Try This

5. Trung had $15 in his piggy bank. After his birthday, he had $60 in his bank. How much money did Trung get as birthday presents?

   Answer the question: _____

   Number model: _____

   Check: How do you know your answer makes sense?

   _____

6. Nikhil had $40 in his wallet when he went to the carnival. When he got home, he had $18. How much did he spend at the carnival?

   Answer the question: _____

   Number model: _____

   Check: How do you know your answer makes sense?

   _____

Lesson 2·5 Math Boxes

1. Use addition or subtraction to complete these problems on your calculator.

Enter	Change to	How?
366	66	_____
894	2,894	_____
3,775	3,175	_____
27,581	28,581	_____

SRB 18, 19, 264

2. "What's My Rule?"

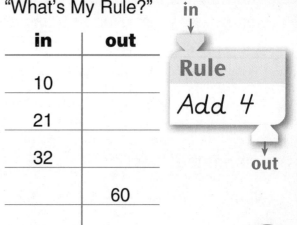

in	out
10	
21	
32	
	60

Rule: Add 4

SRB 203 204

3. 2,345

the 2 means _____

the 3 means _____

the 4 means _____

the 5 means _____

 SRB 18 19

4. Write 5 names in the 120-box.

120

 SRB 14 15

5. Lily had 33 rings in one box and 29 in another. How many did she have in all?

_____ rings

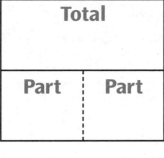

Total	
Part	Part

 SRB 256 257

6. How many squares are shaded? Fill in the oval for the best answer.

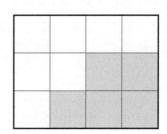

○ 12 ○ 7 ○ 9 ○ 5

forty-one **41**

Date _____ Time _____

LESSON 2·6 Temperature Differences

Use the map on page 220 in the *Student Reference Book* to answer Problems 1–4. Write ? on the diagram for the number you want to find. Write the numbers you know in the comparison diagram. Then solve the problem. Write the answer and a number model.

1. What is the difference between the normal high and low temperatures for San Francisco?

   Answer the question: _____°F

   Number model:

   _____

Quantity

Quantity
   _____
   Difference

   Check: How do you know your answer makes sense?

   _____

   _____

2. What is the difference between the normal high and low temperatures for Seattle?

   Answer the question: _____°F

   Number model:

   _____

Quantity

Quantity
   _____
   Difference

   Check: How do you know your answer makes sense?

   _____

   _____

3. Which city has the *largest* difference between the normal high and low temperatures?

   _____ What is the difference? _____°F

4. Which city has the *smallest* difference between the normal high and low temperatures?

   _____ What is the difference? _____°F

42   forty-two

LESSON 2·6

National High/Low Temperatures Project

Date	Highest Temperature (maximum)		Lowest Temperature (minimum)		Difference (range)
	Place	Temperature	Place	Temperature	
		°F		°F	°F
		°F		°F	°F
		°F		°F	°F
		°F		°F	°F
		°F		°F	°F
		°F		°F	°F
		°F		°F	°F
		°F		°F	°F
		°F		°F	°F
		°F		°F	°F
		°F		°F	°F
		°F		°F	°F
		°F		°F	°F
		°F		°F	°F
		°F		°F	°F
		°F		°F	°F
		°F		°F	°F
		°F		°F	°F
		°F		°F	°F
		°F		°F	°F

forty-three

Lesson 2·6

Math Boxes

1. Complete the fact extensions.

13 = 6 + 7

_____ = 16 + 7

_____ = 26 + 7

_____ = 106 + 7

_____ = 136 + 7

Unit

2. Fill in the blanks.

_____ + 9 = 50

24 + _____ = 30

_____ = 70 − 8

_____ + 73 = 80

Unit

3. About what time is it? Fill in the circle next to the best answer.

○ **A.** 5:30   ○ **B.** 5:05
○ **C.** 6:25   ○ **D.** 5:25

4. Find the rule and complete the table.

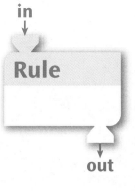

in	out
117	112
119	
	116
	131

5. A vendor sells about 800 ice-cream bars every day. About how many ice-cream bars does the vendor sell in 2 days?

_____ ice-cream bars

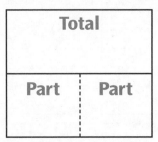

6. Measure the line segment to the nearest centimeter.

_____ cm

_____

Addition Methods

LESSON 2·7

Make a ballpark estimate. Write a number model to show your estimate. Choose at least two problems to solve using the partial-sums method and show your work. You may choose any method you wish to solve the other problems.

Unit: miles

Example:
Ballpark estimate:

300 + 400 = 700

Partial-Sums Method

```
       100s  10s  1s
         3    2   9
    +    4    1   8
         7    0   0
              3   0
    +         1   7
         7    4   7
```

1. Ballpark estimate: _____

```
    43
  + 26
```

2. Ballpark estimate: _____

```
    90
  + 37
```

3. Ballpark estimate: _____

```
   378
  + 401
```

4. Ballpark estimate: _____

```
   172
  + 109
```

5. Ballpark estimate: _____

```
    87
  + 113
```

forty-five 45

Lesson 2·7 Math Boxes

1. Use addition or subtraction to complete these problems on your calculator.

Enter	Change to	How?
173	873	_____
4,501	1,501	_____
5,604	6,604	_____
9,646	9,346	_____

2. "What's My Rule?"

in	out
4	
	12
0	
	21

   Rule: Add 9

   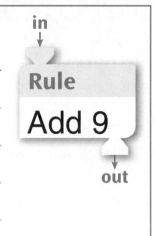

3. In 1,532,

   the 1 means _____

   the 5 means _____

   the 3 means _____

   the 2 means _____

4. Write at least 5 names for 1,000.

   1,000

5. Austin read his book for 45 minutes on Monday and for 25 minutes on Tuesday. How many more minutes did he read on Monday?

   Quantity [_____]

   Quantity [_____]
   Difference

   _____ minutes

6. How long is the fence around the flowers?

   _____ feet

   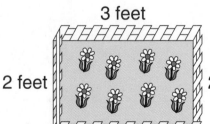

46 forty-six

| Date | Time |

LESSON 2·8 Subtraction Methods

Make a ballpark estimate. Write a number model to show your estimate. Choose at least two problems to solve using the counting-up method and show your work. You may choose any method you wish to solve the other problems.

Unit
lunches

Example:
Ballpark estimate:
$230 - 200 = 30$

```
  234
-  187
```

Counting-Up Method

```
  187
+   ③
  190
+  ⑩
  200
+  ㉚
  230
+   ④
  234
```

$3 + 10 + 30 + 4 = 47$

1. Ballpark estimate:
_____

```
   63
-  37
```

2. Ballpark estimate:
_____

```
   91
-  46
```

3. Ballpark estimate:
_____

```
  129
-  112
```

4. Ballpark estimate:
_____

```
  283
-  256
```

5. Ballpark estimate:
_____

```
  752
-  387
```

forty-seven **47**

Date   **Time**

LESSON 2·8   Name-Collection Boxes

1. Three names do not belong. Mark them with a big **X**.

   100

   1,680 − 1,580

   25 + 25 + 25   80
                         +30
   30 + 70   ─────

     63   1,000
   +37   −100
   ──────   ──────

   2 fifties   9,999
                   −9,899
   48 + 52   ──────

2. Write at least 10 names for 40.

   40

3. Write at least 10 names for 200.

   200

4. Write at least 10 names for 1,000.

   1,000

48   forty-eight

Date Time

Lesson 2·8 Math Boxes

1. Complete the fact extensions.

   Unit

   6 + 5 = 11

   16 + 5 = _____

   26 + 5 = _____

   86 + 5 = _____

   126 + 5 = _____

2. Fill in the blanks.

   Unit
   days

   _____ + 53 = 60

   90 = 3 + _____

   132 = 140 − _____

   198 + _____ = 210

3. What time does the clock show?

   _____

   What time will it be in 30 minutes?

   _____

4. Find the rule and complete the table.

in	out
102	122
130	
	184
	193
188	

5. Corey had $75. He bought a new baseball for $18. How much money does he have now?

   _____

   Number model:

   _____

   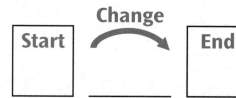

6. Measure the line segment to the nearest $\frac{1}{2}$ centimeter.

   _____ cm

   _____

forty-nine 49

Lesson 2·8 Subtraction Methods

Make a ballpark estimate. Write a number model to show your estimate. Choose at least two problems to solve using the trade-first method and show your work. You may choose any method you wish to solve the other problems.

Unit
dollars

Example:
Ballpark estimate:
$250 - 200 = 50$

Trade-First Method

100s	10s	1s
1	14	
2̷	4̷	7
−1	8	6
	6	1

1. Ballpark estimate: _____

    74
  − 29

2. Ballpark estimate: _____

    96
  − 37

3. Ballpark estimate: _____

    208
  − 106

4. Ballpark estimate: _____

    271
  − 248

5. Ballpark estimate: _____

    826
  − 172

Date _____ Time _____

LESSON 2·9 Number Stories with Several Addends

1. José bought milk for 35 cents, apple juice for 55 cents, grape juice for 45 cents, and orange juice for 65 cents. How much money did he spend?

   Answer the question: _____
   (unit)

   Number model:

   _____

Total			
Part	Part	Part	Part

   Check: How do you know your answer makes sense?

   _____
   _____

2. Michelle drove from Houston, Texas to Wichita, Kansas. On the first day, she drove 245 miles. On the second day, she drove 207 miles. On the third day, she drove 158 miles and arrived in Wichita. How many miles did she drive in all?

   Answer the question: _____
   (unit)

   Number model:

   _____

Total		
Part	Part	Part

   Check: How do you know your answer makes sense?

   _____
   _____

fifty-one **51**

LESSON 2·9 **Number Stories with Several Addends** *continued*

3. Zookeepers watched a clutch of 54 python eggs. On the first day, 18 eggs hatched. On the next day, 11 more hatched. How many eggs did not hatch?

   Answer the question: _____
                             (unit)

Total		
Part	Part	Part

   Number model:
   _____

   Check: How do you know your answer makes sense?

   _____
   _____
   _____

4. Carl has $2.50 for juice or milk at lunch. On each of 2 days, he buys grape juice for 45 cents. On the third day, he buys milk for 40 cents. How much money does he have left?

   Answer the question: _____
                             (unit)

Total			
Part	Part	Part	Part

   Number model:
   _____

   Check: How do you know your answer makes sense?

   _____
   _____
   _____

fifty-two

Date _____ Time _____

LESSON 2·9 Math Boxes

1. Use addition or subtraction to solve these problems on your calculator.

Enter	Change to	How?
409	909	_____
3,291	291	_____
10,538	10,138	_____
12,201	15,201	_____

2. "What's My Rule?"

in	out
14	
24	
39	
	42
	65

   Rule: Subtract 7

3. In 5,639, the 5 means _____.
   Fill in the circle next to the best answer.

   Ⓐ 500
   Ⓑ 5,000
   Ⓒ 50
   Ⓓ 5

4. Fill in the tag. Write at least 5 names for that number.

5. Jack answered 29 questions. José answered 37 questions. How many fewer questions did Jack answer than José?

    Quantity

   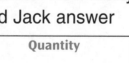 Quantity

   Difference

   ____ questions

6. Which tool would you use to measure the following?

   | yardstick | ruler | thermometer |

   Temperature _____

   Height of the ceiling _____

   Length of your thumb _____

fifty-three 53

Date Time

LESSON 2·10 Math Boxes

1. Which tool would you use to measure the following items:

   | meterstick | 6 in. ruler | thermometer |

   Outdoor temperature _____

   Length of your calculator _____

   Height of the door _____

2. Circle the best unit of measurement.

   Distance to the Galápagos Islands
   kilometers   centimeters   meters

   Width of your thumbnail
   kilometers   centimeters   meters

   Length of your *Student Reference Book*
   kilometers   centimeters   meters

3. Measure the line segment to the nearest $\frac{1}{2}$ inch.

   _____ in.

   _____

4. Measure the line segment to the nearest $\frac{1}{2}$ centimeter.

   _____ cm

   _____

5. How many squares are shaded?

   _____ squares

   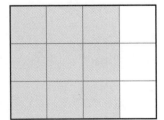

6. How long is the fence around the house?

   _____ meters

   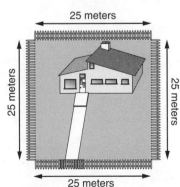

54   fifty-four

LESSON	Estimating and Measuring Lengths
3·1	

Work with a partner. Estimate the lengths of things in the classroom in "class shoe" units. Write the estimate. Then use the class shoe strip to measure the object. Write the measurement.

1.

Object	Estimate	Measurement
	about _____ class shoes	about _____ class shoes
	about _____ class shoes	about _____ class shoes
	about _____ class shoes	about _____ class shoes
	about _____ class shoes	about _____ class shoes
	about _____ class shoes	about _____ class shoes
	about _____ class shoes	about _____ class shoes
	about _____ class shoes	about _____ class shoes
	about _____ class shoes	about _____ class shoes

2. Why is it important to use the same unit everyone else is using to measure things?

_____

_____

_____

Date _____ Time _____

LESSON 3·1 — Addition and Subtraction Practice

Make a ballpark estimate to use to check your answer. Write a number model for your estimate. Add or subtract.

Unit
pumpkin seeds

1. Ballpark estimate:

    _____

    ```
      681
    + 253
    ```

2. Ballpark estimate:

    _____

    ```
      749
    + 161
    ```

3. Ballpark estimate:

    _____

    ```
      417
    + 386
    ```

4. Ballpark estimate:

    _____

    ```
      472
    - 253
    ```

5. Ballpark estimate:

    _____

    ```
      728
    - 173
    ```

6. Ballpark estimate:

    _____

    ```
      550
    - 364
    ```

56   fifty-six

Date _____ Time _____

LESSON 3·1 Math Boxes

1. Add.

    Unit

    9 + 22 + 11 = _____

    13 + 17 + 16 = _____

    24 + 6 + 9 = _____

2. Use addition or subtraction to complete these problems on your calculator.

Enter	Change to	How?
141	191	_____
406	906	_____
1,873	1,273	_____
1,462	5,462	_____

3. Order these numbers from smallest to largest.

    1,060 _____

    1,600 _____

    1,006 _____

    6,001 _____

4. 148 + 45 = _____

    Unit

    Make a ballpark estimate. Write your number model.

    _____ + _____ = _____

5. Solve using the counting-up method. Show your work.

    Unit

       42
    − 14
    ____

6. Circle the event that is *unlikely* to happen.

    If you toss a coin 20 times, it will always land on HEADS.

    If you toss a coin 20 times, it will land on HEADS some of the times.

fifty-seven **57**

LESSON 3·2
Measuring Line Segments

1. Use Ruler A to measure to the nearest inch (in.).
   Use Ruler D to measure to the nearest centimeter (cm).

   Ruler A   **Ruler D**

   about ___ in.   about ___ cm

   about ___ in.   about ___ cm

   about ___ in.   about ___ cm

2. Use Ruler B to measure to the nearest $\frac{1}{2}$ inch (in.).
   Use Ruler D to measure to the nearest $\frac{1}{2}$ centimeter (cm).

   Ruler B   **Ruler D**

   about ___ in.   about ___ cm

   about ___ in.   about ___ cm

   about ___ in.   about ___ cm

Try This

3. Use Ruler C to measure to the nearest $\frac{1}{4}$ inch (in.).
   Use Ruler D to measure to the nearest millimeter (mm).

   Ruler C   **Ruler D**

   about ___ in.   about ___ mm

   about ___ in.   about ___ mm

   about ___ in.   about ___ mm

Date Time

Lesson 3·2 Math Boxes

1. Count by 6s.

   _57_, ____, ____, ____,

   _81_, ____, ____, ____,

   ____, ____, ____, ____,

   ____, ____, ____, ____

2. Measure to the nearest $\frac{1}{4}$ inch.

   about ____ in.

   _____

   Draw a line segment $1\frac{1}{2}$ inches long.

3. Write <, >, or =.

   69 ____ 96

   101 ____ 110

   2¢ ____ 5¢

   1,000 ____ 999

4. Pamela had $38. She spent ____ on shoes. She has $15 left.

   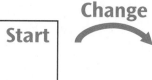

5. Book Club Totals

   Number of Children

   ```
              X
           X  X
        X  X  X
     X  X  X  X
     X  X  X  X
     X  X  X  X
     +--+--+--+--+
     0  1  2  3  4
        Books Read
   ```

   Maximum number of books read:

   _____

6. Courtney has 8 pennies. She shares them equally with Nicholas. How many pennies do they each get? Fill in the circle for the best answer.

   ○ A. 16 pennies

   ○ B. 8 pennies

   ○ C. 4 pennies

   ○ D. 2 pennies

fifty-nine **59**

Date _____ Time _____

LESSON 3·3 Measures Hunt

Find out about how long some objects are.
These objects will be **personal references.**
Use your personal references to estimate the lengths of other things.

1. Find things that are about 1 inch long, 1 foot long, and 1 yard long.
   Use a ruler, tape measure, or yardstick.
   List your objects below.

About 1 inch (in.)	About 1 foot (ft)	About 1 yard (yd)
_____	_____	_____
_____	_____	_____
_____	_____	_____
_____	_____	_____
_____	_____	_____

2. Find things that are about 1 centimeter long, 1 decimeter long, and 1 meter long.
   Use a ruler, tape measure, or meterstick.
   List your objects below.

About 1 centimeter (cm)	About 1 decimeter (dm)	About 1 meter (m)
_____	_____	_____
_____	_____	_____
_____	_____	_____
_____	_____	_____
_____	_____	_____

LESSON 3·3 Estimating Lengths

1. Follow these steps using **U.S. customary** units: inches (in.), feet (ft), or yards (yd). Then follow these steps using **metric** units: millimeters (mm), centimeters (cm), decimeters (dm), or meters (m).

    ◆ Use personal references to estimate the measures.

    ◆ Record your estimates. Be sure to write the units.

    ◆ Measure with a ruler or tape measure. Record your measurements.

Objects	U.S. Customary Units		Metric Units	
	Estimate	Measurement	Estimate	Measurement
height of your desk				
long side of your calculator				
short side of the classroom				
distance around your head				

2. Choose your own objects to estimate and measure.

Objects	U.S. Customary Units		Metric Units	
	Estimate	Measurement	Estimate	Measurement

sixty-one 61

Lesson 3·3 Math Boxes

1. 8 + 6 = _____

   8 + 6 + 7 = _____

   8 + 6 + 7 + 5 = _____

   ```
     17        17        17
   +  8         8         8
              + 5         5
                        + 19
   ```

   Unit

2. Use addition or subtraction to complete these problems on your calculator.

Enter	Change to	How?
267	307	_____
1,039	539	_____
1,374	1,874	_____
15,866	11,866	_____

3. Order these numbers from largest to smallest.

   1,164 _____

   1,104 _____

   1,146 _____

   1,416 _____

4. Estimate.
   Is 82 − 49 closer to 40 or to 30?

   _____

   Show the number model you used.

   Unit

   _____ − _____ = _____

5. Solve using the trade-first method. Show your work.

   ```
     66
   − 38
   ```

   Unit

6. Fill in the circle for the best answer.

   If you toss a quarter 100 times, you can be certain it will

   Ⓐ always land on HEADS.

   Ⓑ always land on either HEADS or TAILS.

   Ⓒ land on HEADS 99 times.

   Ⓓ land on HEADS exactly $\frac{1}{2}$ of the time.

LESSON 3·4 Perimeters of Polygons

1. Record the **perimeter** (the distance around) of your straw rectangle and parallelogram.

    rectangle: about _____ inches   parallelogram: about _____ inches

2. Use a tape measure to find each side and the perimeter.

Polygon	Each Side	Perimeter
triangle	about _____ in., about _____ in., about _____ in.	about _____ in.
triangle	about _____ in., about _____ in., about _____ in.	about _____ in.
square	about _____ in.	about _____ in.
rhombus	about _____ in.	about _____ in.
trapezoid	about _____ in., about _____ in. about _____ in., about _____ in.	about _____ in.

3. Find the perimeter, in inches, of the figures below.

   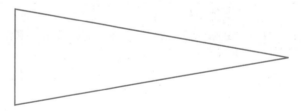

   _____   _____

Try This

4. Draw each shape on the centimeter grid.

   square with perimeter = 16 cm   rectangle with perimeter = 20 cm

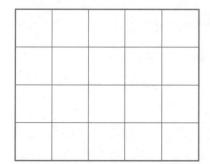   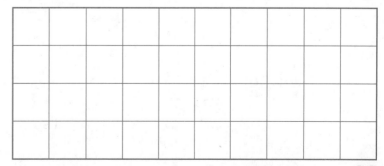

sixty-three **63**

Date Time

LESSON 3·4 Body Measures

Work with a partner to find each measurement to the nearest $\frac{1}{2}$ inch.

	Adult at Home	Me (Now)	Me (Later)
Date	_____	_____	_____
height	about ____ in.	about ____ in.	about ____ in.
shoe length	about ____ in.	about ____ in.	about ____ in.
around neck	about ____ in.	about ____ in.	about ____ in.
around wrist	about ____ in.	about ____ in.	about ____ in.
waist to floor	about ____ in.	about ____ in.	about ____ in.
forearm	about ____ in.	about ____ in.	about ____ in.
hand span	about ____ in.	about ____ in.	about ____ in.
arm span	about ____ in.	about ____ in.	about ____ in.
_____	about ____ in.	about ____ in	about ____ in.
_____	about ____ in.	about ____ in.	about ____ in.
_____	about ____ in.	about ____ in.	about ____ in.

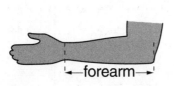
forearm

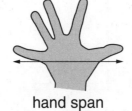
hand span

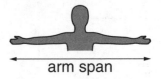
arm span

Lesson 3·4 Math Boxes

1. Count back by 7s.

   _____, 98, _____, _____,

   _____, _____, 63, _____,

   _____, _____, _____, _____,

   _____, _____, _____, _____

2. Measure to the nearest centimeter.

   about _____ cm

   _____

   Draw a line segment 4 centimeters long.

3. Write <, >, or =.

   1,069 _____ 10,691

   6,589 _____ 6,859

   42,617 _____ 42,429

   Make up your own.

   _____ _____ _____

4. 53 people were standing in line at 9:00 A.M. 97 people were standing in line at 10:00 A.M. How many more people were standing in line at 10:00 A.M.?

   _____ people

   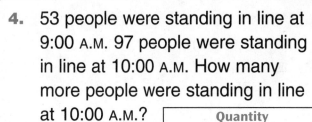
   Difference

5. Shade to show the following data:
   A is 4 cm.
   B is 3 cm.
   C is 8 cm.
   D is 7 cm.

   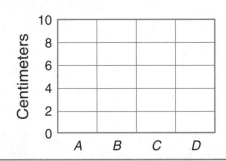

6. 2 children share 12 toys equally. How many toys does each child get?

   _____ toys

sixty-five **65**

Date _____ Time _____

LESSON 3·5 Math Boxes

1. Measure to the nearest $\frac{1}{2}$ inch. Fill in the oval next to the best answer.

   ○ 1 in.

   ○ $1\frac{1}{2}$ in.

   ○ 2 in.

   ○ $2\frac{1}{2}$ in.

2. What is the perimeter?

   3 cm, 4 cm, 2 cm, 2 cm, 4 cm, 3.5 cm

   _____ (unit)

3. Write <, >, or =. Use a tape measure to help.

   $1\frac{1}{2}$ feet _____ 16 inches

   3 feet _____ 2 yards

   5 feet _____ 60 inches

   55 inches _____ 1 yard

4. Add. Show your work.

   Ballpark estimate:

   _____

   ```
     555
   + 192
   ```

5. Solve.

   9 + 1 + 4 = _____

   _____ = 3 + 7 + 8 + 2

   3 + 15 + 7 + 4 = _____

6. Solve.

   3 × 0 = _____

   _____ = 5 × 0

   0 × 7 = _____

   9 × 0 = _____

LESSON 3·6 Geoboard Perimeters

Materials ☐ geoboard and rubber bands or geoboard dot paper

Work with a partner.

1. Suppose that the distance between two pins is 1 unit. Make a rectangle with a perimeter of 14 units. Use rubber bands and a geoboard, or draw the rectangle on dot paper. Record the lengths of the sides in the table.

2. Now make a different rectangle that also has a perimeter of 14 units. Record the lengths of the sides for this shape.

3. Complete the table for other perimeters.

4. Try to make a rectangle or square with a perimeter of 13 units.

5. Try to make other rectangles or squares with perimeters that are an odd number of units.

Geoboard Perimeters

Perimeter	Longer sides	Shorter sides
14 units	____ units	____ units
14 units	____ units	____ units
14 units	____ units	____ units
12 units	____ units	____ units
12 units	____ units	____ units
12 units	____ units	____ units
16 units	____ units	____ units
16 units	____ units	____ units
16 units	____ units	____ units
16 units	____ units	____ units

Try This

Change the unit. Now 1 unit is double the distance between two points. Make a rectangle or square whose perimeter is an odd number of units.

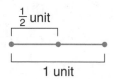

Follow-Up

Look for a pattern in your table. Can you find one? Now, without using a geoboard or dot paper, find the lengths of the sides of a rectangle or square with a perimeter of 24 units. Then make or draw the shape to check your answer.

sixty-seven **67**

Date _____ Time _____

LESSON 3·6 Tiling with Pattern Blocks

Materials ☐ pattern blocks: square, triangle, narrow rhombus
☐ crayons

Work with a partner.

1. Use square pattern blocks. Look at the top rectangle on the next page. Cover as much of the rectangle as you can, placing all of the blocks inside it. There may be uncovered spaces at the edges. Do not overlap the blocks. Line them up so that there are no gaps. This is called **tiling**.

2. Count and record the number of blocks you used.

3. Trace around the edges of each block. Then color any spaces not covered by blocks. Estimate how many blocks would be needed to cover the colored spaces.

4. Record how many blocks are needed to cover the whole rectangle.

5. Tile the second rectangle with triangles. Repeat Steps 2–4 above.

6. Tile the third rectangle with narrow rhombuses. Repeat Steps 2–4 above.

Follow-Up

7. The **area** of a shape is a measure of the space inside the shape. You measured the area of a rectangle three ways: with squares, triangles, and narrow rhombuses. Record the areas below.

    The area of the rectangle is about _____ squares.

    The area of the rectangle is about _____ triangles.

    The area of the rectangle is about _____ narrow rhombuses.

8. Which of the three pattern blocks has the largest area? _____

    Which has the smallest area? _____

    How did you decide? _____

    _____

LESSON 3·6
Tiling with Pattern Blocks *continued*

Cover this rectangle with squares.

About _____ squares cover the whole rectangle.

Cover this rectangle with triangles.

About _____ triangles cover the whole rectangle.

Cover this rectangle with narrow rhombuses.

About _____ narrow rhombuses cover the whole rectangle.

sixty-nine

LESSON 3·6 Straw Triangles

Materials
☐ 4-inch, 6-inch, and 8-inch straws
☐ twist-ties

Work in a group to make as many different-size triangles as you can out of the straws and twist-ties. (Be sure that straws are touching at all ends.) Before you start, decide how you will share the work.

For each triangle, record the length of each side and the perimeter in the chart. The triangle made out of the shortest straws is already recorded.

Straw Triangles

Side 1	Side 2	Side 3	Perimeter
4 in.	4 in.	4 in.	12 in.

Follow-Up
Discuss these questions with others in your group.

1. Which triangles have three equal sides?

   _____

2. Which pairs of triangles have the same perimeter?

   _____

3. By looking at your constructions, estimate which triangle of each pair of triangles in Problem 2 has the larger area (space inside the triangles).

   _____

4. What happens if you try to make a triangle out of two 4-inch straws and one 8-inch straw?

   _____

seventy

Date _____ Time _____

LESSON 3·6 Math Boxes

1. Which tool would you use to measure the following?

   | yardstick | ruler | thermometer |

   temperature _____

   height of the ceiling _____

   length of your thumb _____

2. On the centimeter grid below, draw a shape with an area of 12 square centimeters.

   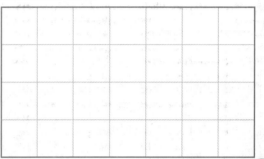

3. Justin bought 2 gallons of milk. Each gallon cost $2.79. He paid with a $10 bill. How much change did he receive? _____

4. Sarah had $0.54. She found some coins. Now she has $0.83. How much money did she find?

   _____

   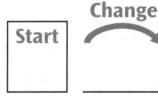

5. Fill in the circle next to the best answer.

   ○ A. It is certain to be sunny tomorrow.

   ○ B. A tossed quarter will land on either HEADS or TAILS.

   ○ C. A rolled die will always land on 6.

   ○ D. The sun will set at the same time it did last month.

6. Fill in the empty frames.

   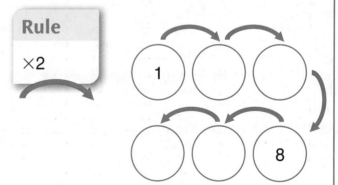

seventy-one **71**

Date _____ Time _____

LESSON 3·7 Areas of Rectangles

Draw each rectangle on the grid. Make a dot inside each small square in your rectangle.

1. Draw a 3-by-5 rectangle.

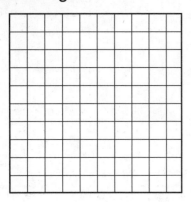

Area = _____ square units

2. Draw a 6-by-8 rectangle.

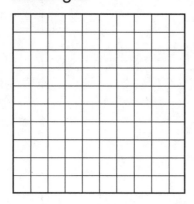

Area = _____ square units

3. Draw a 9-by-5 rectangle.

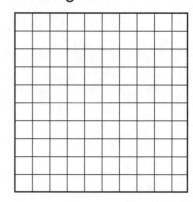

Area = _____ square units

Fill in the blanks.

4.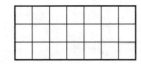

This is a _____-by-_____ rectangle.

Area = _____ square units

5.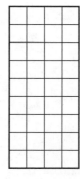

This is a _____-by-_____ rectangle.

Area = _____ square units

6.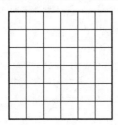

This is a _____-by-_____ rectangle.

Area = _____ square units

7.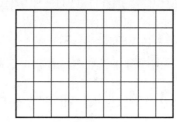

This is a _____-by-_____ rectangle.

Area = _____ square units

Lesson 3·7 — **Math Boxes**

1. Measure to the nearest centimeter.

   about _____ cm

   _____

   Draw a line segment 4 centimeters long.

2. Draw a shape with an area of 9 square centimeters.

   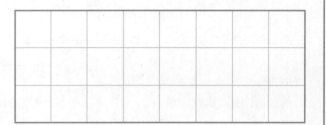

3. Write the equivalent lengths. Use a tape measure to help.

   3 yards = _____ feet

   _____ inches = 2 yards

   50 millimeters = _____ centimeters

   3 meters = _____ centimeters

4. Subtract. Show your work.

   Ballpark estimate:

   _____

   943
   −409

5. Solve.

   8 + 3 + 2 + 2 = _____

   _____ = 9 + 14 + 1 + 2

   4 + 3 + 11 + 6 = _____

   85 + 16 + 4 + 15 = _____

6. Solve.

   1 × 2 = _____

   _____ = 1 × 4

   5 × 1 = _____

   8 × 1 = _____

seventy-three **73**

LESSON 3·8 More Areas of Rectangles

Make a dot inside each small square in one row. Then fill in the blanks.

1.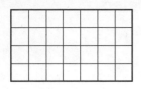

2.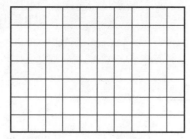

3.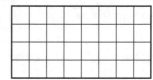

Number of rows: ____

Squares in a row: ____

Area = ____ square units

Number Model:

____ × ____ = ____

Number of rows: ____

Squares in a row: ____

Area = ____ square units

Number Model:

____ × ____ = ____

Number of rows: ____

Squares in a row: ____

Area = ____ square units

Number Model:

____ × ____ = ____

Now draw the rectangle on the grid. Then fill in the blanks.

4. Draw a 5-by-7 rectangle.

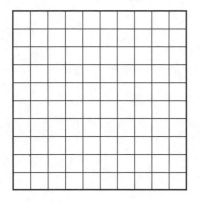

Area = ____ square units

Number Model:

____ × ____ = ____

5. Draw an 8-by-8 rectangle.

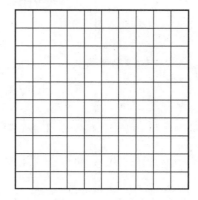

Area = ____ square units

Number Model:

____ × ____ = ____

6. Draw a 3-by-9 rectangle.

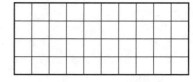

Area = ____ square units

Number Model:

____ × ____ = ____

Date　　　　　　　　　　Time

Lesson 3·8　Math Boxes

1. Circle the best unit of measurement.

distance to Spain

　miles　　centimeters　　inches

width of a crayon

　miles　　centimeters　　feet

length of your journal

　miles　　yards　　inches

2. Find the perimeter.

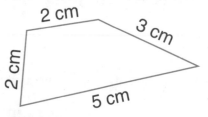

Fill in the circle for the best answer.
- A. 13 cm
- B. 12 cm
- C. 8 cm
- D. 4 cm

3. When I left home, I had $4.00. I spent $0.73 at the fruit stand and $2.59 at the grocery store. How much did I spend in all?

_____

How much do I have left?

_____

4. Fill in the unit box. Write the missing number in the diagram.

Unit

+107　　End 392

Write a number model.

_____ + _____ = _____

5. Describe 2 events that you are certain *will not* happen today.

_____

_____

_____

6. Fill in the empty frames.

Rule
×0

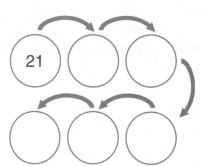

21

seventy-five　**75**

Date _____ Time _____

LESSON 3·9 Diameters and Circumferences

1. Find numbers on the label of your can. Write some of them below. Also write the unit if there is one.

   _____

2. Record the diameter and circumference of your can.

   can letter: _____   **diameter:** about _____ cm

   circumference: about _____ cm

3. Write the rule linking diameter and circumference:

   _____

   _____

4. Fill in the empty frames. Use two rules.

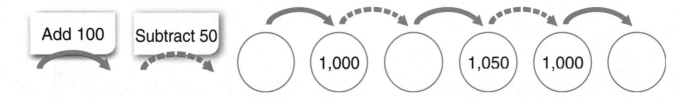

5.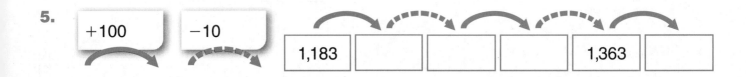

6.

76   seventy-six

Date _____ Time _____

LESSON 3·9 Math Boxes

1. Measure to the nearest centimeter.

    about _____ cm

    _____

    Draw a line segment 7 centimeters long.

2.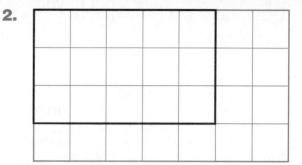

    Area: _____ square cm

3. Write <, >, or =. You may wish to use a tape measure.

    6 decimeters _____ 60 millimeters

    3 yards _____ 36 inches

    2 centimeters _____ 4 meters

    Write your own.

    _____

4. Add. Show your work.

    Ballpark estimate:

    _____ + _____ = _____

    $$\begin{array}{r} 8{,}916 \\ +\ 7{,}504 \\ \hline \end{array}$$

5. Write <, >, or =.

    4 + 5 + 6 _____ 3 + 5 + 7

    7 + 9 + 5 _____ 6 + 6 + 8

    2 + 11 + 4 _____ 7 + 1 + 9

    15 + 7 + 5 _____ 9 + 9 + 9

    4 + 5 + 6 _____ 3 + 7 + 6

6. Solve.

    2 × 2 = _____

    2 × 3 = _____

    _____ = 2 × 4

    2 × 6 = _____

seventy-seven **77**

Date _____ Time _____

LESSON 3·10 Math Boxes

1. Describe 2 events that are *impossible*.

   _____

   _____

   _____

2. Karan shared 18 jelly beans equally with her sister Sonia. How many jelly beans did they each get? Draw a picture or an array.

   _____ jelly beans

3. Solve.

   $4 \times 0 =$ _____

   $0 \times 3 =$ _____

   _____ $= 8 \times 0$

   _____ $= 0 \times 10$

4. Solve.

   $3 \times 1 =$ _____

   _____ $= 1 \times 7$

   _____ $= 4 \times 1$

   $1 \times 6 =$ _____

5. Solve.

   $2 \times 5 =$ _____

   $1 \times 2 =$ _____

   _____ $= 10 \times 2$

   _____ $= 2 \times 7$

6. Fill in the empty frames.

   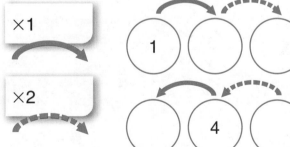

Date _____ Time _____

LESSON 4·1 — Solving Multiplication Number Stories

Use the Variety Store Poster on page 215 of the *Student Reference Book*.

For each number story:

♦ Fill in a multiplication/division diagram. Write ? for the number you need to find. Write the numbers you already know.

♦ Use counters or draw pictures to help you find the answer.

♦ Record the answer with its unit. Check whether your answer makes sense.

1. Yosh has 4 boxes of mini stock cars. There are 10 stock cars in each box. How many stock cars does he have?

boxes	cars per box	cars in all

   Answer: _____
              (unit)

   How do you know your answer makes sense? _____

   _____

2. There are 100 file cards in each package. How many cards are in 5 packages of file cards?

packages	cards per package	cards in all

   Answer: _____
              (unit)

   How do you know your answer makes sense? _____

   _____

3. Use a separate sheet of paper. Write your own multiplication number story. Write how you know your answer makes sense.

seventy-nine **79**

Date _____ Time _____

LESSON 4·1 Math Boxes

1. Each square equals 1 sq cm. Find the area.

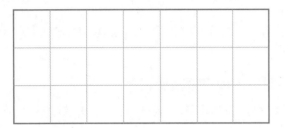

Area: ____ square centimeters

2. Scientists counted 136 eggs in a clutch of turtle eggs. 87 eggs did *not* hatch.
Estimate how many eggs hatched.

About _____

Number model for estimate:

_____

3. Make each sentence true. Use >, <, or =.

8 + 6 ____ 7 + 7

17 − 9 ____ 9 + 8

14 − 5 ____ 11 − 4

4. Find the perimeter.

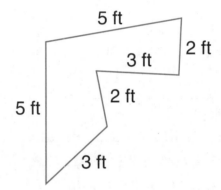

Perimeter = _____
(unit)

5. How many rows of dots?   . . . . .
. . . . .

_____

How many dots in each row?

_____

How many dots in all? _____

6. Solve.

45,582 − 100 = _____

45,582 + 100 = _____

45,582 + 1,000 = _____

45,582 − 1,000 = _____

80   eighty

Date _____ Time _____

LESSON 4·2 More Multiplication Number Stories

- Fill in the multiplication/division diagram.
- Make an array with counters. Mark the dots to show the array.
- Find the answer. Write the unit with your answer. Write a number model.

1. Mrs. Kwan has 3 boxes of scented markers. Each box has 8 markers. How many markers does she have?

boxes	markers per box	markers in all

Answer: _____ (unit)

Number model: _____

2. Monica keeps her doll collection in a case with 5 shelves. On each shelf there are 6 dolls. How many dolls are in Monica's collection?

shelves	dolls per shelf	dolls in all

Answer: _____ (unit)

Number model: _____

3. During the summer Jack mows lawns. He can mow 4 lawns per day. How many lawns can he mow in 7 days?

days	lawns per day	lawns in all

Answer: _____ (unit)

Number model: _____

eighty-one 81

Date　　　　　　　　　　　　Time

LESSON 4·2 Measuring Perimeter

Measure the perimeter of each figure in inches.

1.

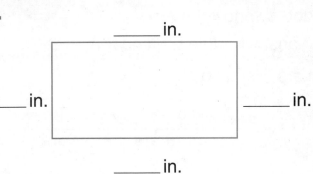

_____ in.
_____ in. _____ in.
_____ in.

Perimeter: _____ inches

2.

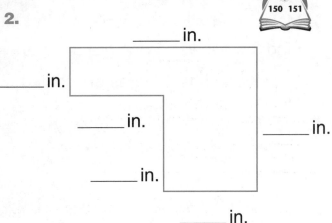

_____ in.
_____ in.
_____ in. _____ in.
_____ in.
_____ in.

Perimeter: _____ inches

3.

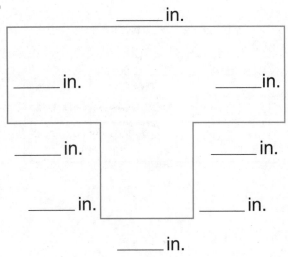

_____ in.
_____ in. _____ in.
_____ in. _____ in.
_____ in. _____ in.
_____ in.

Perimeter: _____ inches

4. **Try This**

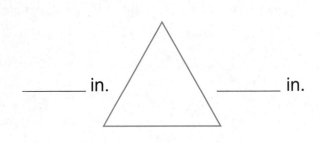

_____ in. _____ in.
_____ in.

Perimeter: _____ inches

5. Draw any figure with a perimeter of 20 centimeters.

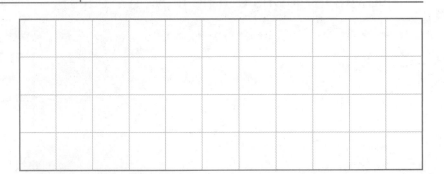

Lesson 4·2 Math Boxes

1. Complete the bar graph.

   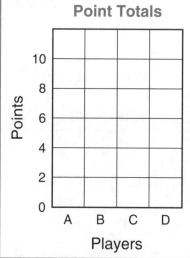

   Player A scores 4 points.
   Player B scores 8 points.
   Player C scores 3 points.
   Player D scores 9 points.

   SRB 86 87

2. 10 packs of gum on the shelf in the candy store. 8 sticks of gum per pack. How many sticks of gum in all?

packs	sticks of gum per pack	sticks of gum in all

   Answer: _____

   SRB 259

3. Solve. Make a ballpark estimate to check that the answer makes sense.

   Unit

   _____ = 648 + 209

   estimate:

   _____

   SRB 192

4. Solve.

   3 × 5 = _____

   3 nickels = _____ ¢

   _____ = 4 × 5

   _____ ¢ = 4 nickels

   SRB 52 53 56

5. Fill in the empty frames.

   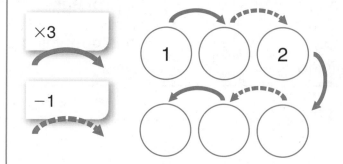

   SRB 200 201

6. 1,798

   Which digit is in the tens place? _____

   Which digit is in the hundreds place? _____

   Which digit is in the ones place? _____

   Which digit is in the thousands place? _____

   SRB 18 19

Date _____ Time _____

LESSON 4·3 Division Practice

Use counters to find the answers. Fill in the blanks.

16¢ shared equally

1. by 2 people:

   _____ ¢ per person

   _____ ¢ remaining

2. by 3 people:

   _____ ¢ per person

   _____ ¢ remaining

3. by 4 people:

   _____ ¢ per person

   _____ ¢ remaining

25¢ shared equally

4. How many people get 5¢?

   _____ people

   _____ ¢ remaining

5. How many people get 3¢?

   _____ people

   _____ ¢ remaining

6. How many people get 6¢?

   _____ people

   _____ ¢ remaining

30 stamps shared equally

7. by 10 people:

   _____ stamps per person

   _____ stamps remaining

8. by 5 people:

   _____ stamps per person

   _____ stamps remaining

9. by 6 people:

   _____ stamps per person

   _____ stamps remaining

10. 21 days
    7 days per week

    _____ weeks

    _____ days remaining

11. 32 crayons
    6 crayons per box

    _____ boxes of crayons

    _____ crayons remaining

12. 24 eggs
    6 eggs per row

    _____ rows of eggs

    _____ eggs remaining

13. There are 18 counters in an array. There are 6 rows.

    How many counters are in each row? _____ counters per row

14. Five children share 12 markers equally. How many markers does each child get? _____ markers with _____ markers remaining

84 eighty-four

Date _____ Time _____

LESSON 4·3 Math Boxes

1. On the centimeter grid below, draw a shape with an area of 10 square centimeters.

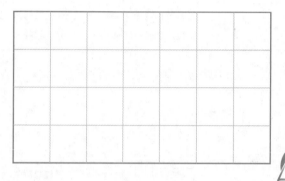

SRB 154–156

2. Corinne wants new tires for her bicycle. They cost $41.10 each, with tax included. Estimate about how much money she will need.

about $ _____

Number model:

_____

SRB 191

3. Use >, <, or =.

Unit

9 + 9 _____ 13 + 5

13 − 4 _____ 11 − 5

11 − 4 _____ 13 − 8

SRB 13 50 51

4. Find the perimeter. Fill in the circle for the best answer.

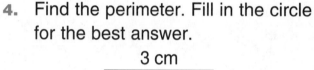
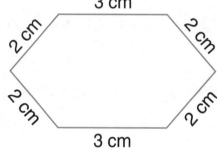

Ⓐ 14 cm   Ⓑ 6 cm

Ⓒ 7 cm   Ⓓ 12 cm

SRB 150 151

5. Complete the number model for the 4 by 4 array.

How many rows? _____

How many dots in each row?

_____

____ × ____ = ____

SRB 64 65

6. Write the number that is 100 more.

76 _____

300 _____

471 _____

8,634 _____

5,925 _____

SRB 18 19

eighty-five **85**

Date _____ Time _____

LESSON 4·4 Solving Multiplication and Division Number Stories

Solve each number story. Use counters or draw an array to help you. Fill in the diagrams and write number models.

1a. Roberto has 3 packages of pencils. There are 12 pencils in each package. How many pencils does Roberto have in all?

Answer: _____ (unit)

packages	pencils per package	pencils in all

Number model: _____

1b. Roberto gives 3 of his pencils to each of his friends. How many friends will get 3 pencils each?

Answer: _____ (unit)

friends	pencils per friend	pencils in all

Number model: _____

2a. A class of 30 children wants to play ball. How many teams can be made with exactly 6 children on each team?

Answer: _____ (unit)

teams	children per team	children in all

Number model: _____

2b. For another game, the same class of 30 children wants to have exactly 4 children on each team. How many teams can they make?

Answer: _____ (unit)

teams	children per team	children in all

Number model: _____

86 eighty-six

Date   Time

LESSON 4·4   **Math Boxes**

1. Complete the bar graph.

   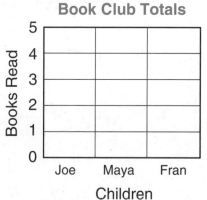

   Joe read 3 books.
   Maya read 2 books.
   Fran read 4 books.

   Median books read: _____ (unit)

2. Show an array and complete a number model to match the diagram.

packs	cards per pack	cards in all
3	6	?

   Number model: _____

3. Add. Make a ballpark estimate.

   Unit

   _____ = 47 + 192

   estimate: _____

   _____ = 147 + 292

   estimate: _____

4. Solve.

   6 × 10 = _____

   6 dimes = _____ ¢

   _____ = 8 × 10

   _____ ¢ = 8 dimes

5. Fill in the empty frames.

   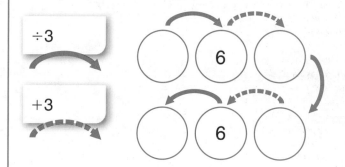

6. What number has

   2 hundreds   5 ones   3 tens

   4 thousands   6 ten-thousands

   Fill in the circle for the best answer and read it to a partner.

   Ⓐ 24,536   Ⓑ 42,356

   Ⓒ 63,542   Ⓓ 64,235

eighty-seven **87**

LESSON 4·5 Subtraction Strategies

Make a ballpark estimate. Write a number model to show your estimate. Choose at least two problems to solve using the counting-up method. You may choose any method you wish to solve the other problems.

1. Ballpark estimate:

   _____

   226
   −134

2. Ballpark estimate:

   _____

   93
   −47

3. Ballpark estimate:

   _____

   487
   −129

4. Ballpark estimate:

   _____

   361
   −248

5. Ballpark estimate:

   _____

   724
   −396

6. Ballpark estimate:

   _____

   515
   −367

Date _____ Time _____

LESSON 4·5   Math Boxes

1. Use the dots to show a 3 × 6 array.

   . . . .
   . . . .
   . . . .
   . . . .
   . . . .

   What is the number model?

   _____ × _____ = _____

2. Maximum number of points scored:

   _____

   Minimum number of points scored:

   _____

   Range of points scored:

   _____

   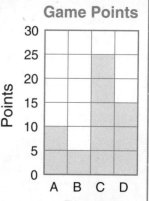

3. Solve. Fill in the oval for the best answer.

   4 rows of chairs

   6 chairs in each row

   How many chairs in all?

   ⬭ 10 chairs   ⬭ 12 chairs

   ⬭ 24 chairs   ⬭ 20 chairs

4. Fill in the number grid.

   | 2,946 | | |

   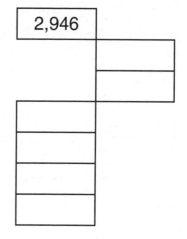

5. Draw a 2 × 4 rectangle.

   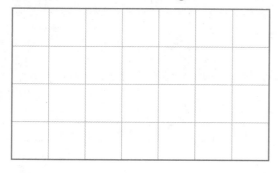

   Number model: ___ × ___ = ___

   Area: ___ square units

6. Which is more?

   $3.45 or $3.09? _____

   $0.34 or $0.09? _____

   $14.50 or $14.55? _____

   $30.15 or $31.05? _____

eighty-nine 89

Date _____ Time _____

Lesson 4·6 Math Boxes

1. Write <, >, or =.

   3 × 2 _____ 2 × 3

   4 × 1 _____ 8 × 0

   5 × 3 _____ 5 × 4

   9 × 0 _____ 0 × 7

   SRB 13 56

2. Write the fact family.

   _____

   _____

   _____

   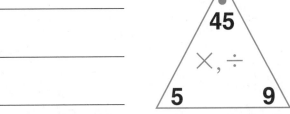

   SRB 55

3. Complete.

   Rule
   ×3

yd	ft
2	
5	
	9
	30

   SRB 52 203 204

4. Use the dots to show a 7-by-6 array.

   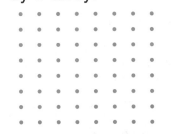

   What is the number model?

   _____ × _____ = _____

   SRB 64 65

5. Lengths (in.) of 13 cats, including the tails:

   30, 29, 28, 24, 29, 35, 16, 27, 29, 36, 28, 31, 32

   What is the maximum length? Fill in the oval for the best answer.

   ◯ 36 inches   ◯ 16 inches

   ◯ 29 inches   ◯ 30 inches

   SRB 79

6. Fill in the empty frames.

   Rule
   +1,000

   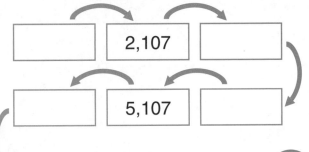

   [___] → [2,107] → [___]

   [___] → [5,107] → [___]

   [___]

   SRB 200 201

90 ninety

Date Time

LESSON 4·7 Math Boxes

1. Draw Xs in a 5-by-9 array.

   How many Xs?

   _____

   Write a number model for the array.

   _____

2. Maximum height of seedlings:

   _____

   Minimum height of seedlings:

   _____

   Range of seedling heights:

   _____

3. Use counters to solve.

   Some children are sharing 22 marbles equally. Each child gets 6 marbles. How many children are sharing?

   _____ (unit)

   How many marbles are left over?

   _____ (unit)

4. Fill in the number grid.

   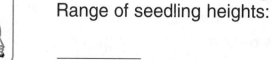

   3,039

5. Draw a shape with an area of 11 square centimeters.

   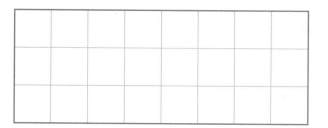

6. Use >, <, or =.

   $40.75 _____ $47.05

   $0.86 _____ $8.00

   $31.02 _____ $31.20

   $107.40 _____ $97.40

ninety-one **91**

Date _____ Time _____

Exploration A: How Many Dots?

Materials ☐ square pattern blocks
☐ calculator

1. Estimate how many dots are in the array at the right.

   About _____ dots

   Make another estimate. Follow these steps.

2. Cover part of the array with a square pattern block. About how many dots does one block cover?

   _____ dots

3. Cover the array. Use as many square pattern blocks as you can. Do not go over the borders of the array. How many blocks did you use?

   _____ blocks

4. Use the information in Steps 2 and 3 to estimate the total number of dots in the array. About _____ dots

Try This

5. Find the exact number of dots in the array. Use a calculator to help you. Total number of dots = _____

Follow-Up

Describe how you found the exact number of dots. _____

92   ninety-two

LESSON 4·8

Exploration B: Setting Up Chairs

1. Record the answer to the problem about setting up chairs from *Math Masters*, page 106.

    There were _____ chairs in the room.

2. Circle dots below to show how you set up the chairs for each of the clues.

Rows of 2	Rows of 3	Rows of 4	Rows of 5
• 1 left over	• 1 left over	• 1 left over	• 0 left over

ninety-three 93

Date _____ Time _____

Lesson 4·8 **Math Boxes**

1. Make equal groups.

   30 days make _____ weeks

   with _____ days left over.

   56 pennies make _____ quarters

   with _____ pennies left over.

2. Write three things that you think are *very likely* to happen.

   _____

   _____

   _____

3. Fill in the circle for the best answer. The perimeter of the square is

   ○ A. 12 cm
   ○ B. 16 cm
   ○ C. 8 cm
   ○ D. 20 cm

   4 cm   4 cm

4. Complete the Fact Triangle. Write the fact family.

   ___ × ___ = ___

   ___ × ___ = ___

   ___ ÷ ___ = ___

   ___ ÷ ___ = ___

   21, 3, ___ ×, ÷

5. 56,937

   Which digit is in the tens place? __3__

   Which digit is in the thousands place? ___

   Which digit is in the hundreds place? ___

   Which digit is in the ones place? ___

6. Use >, <, or =.

   3,065 _____ 3,605

   23,605 _____ 20,365

   32,605 _____ 23,605

   50,007 _____ 50,700

94   ninety-four

LESSON 4·9 Estimating Distances

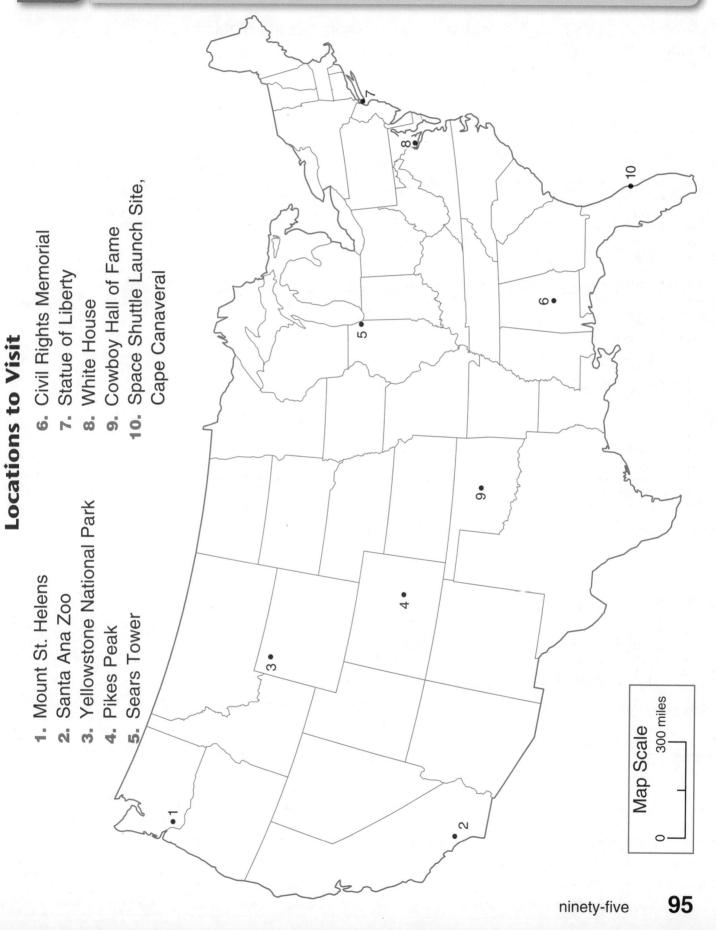

Locations to Visit

1. Mount St. Helens
2. Santa Ana Zoo
3. Yellowstone National Park
4. Pikes Peak
5. Sears Tower
6. Civil Rights Memorial
7. Statue of Liberty
8. White House
9. Cowboy Hall of Fame
10. Space Shuttle Launch Site, Cape Canaveral

Map Scale: 0 — 300 miles

ninety-five 95

LESSON 4·9 A Pretend Trip

Pretend that you want to take a trip to see some of the sights in the United States. Find out about how far it is between locations.

1. Yellowstone National Park is number _____.

   The Cowboy Hall of Fame is number _____.

   The distance between them is about _____ inches on the map.

   That is about _____ miles.

2. Pikes Peak is number _____.

   The White House is number _____.

   The distance between them is about _____ inches on the map.

   That is about _____ miles.

3. The Civil Rights Memorial is number _____.

   Santa Ana Zoo is number _____.

   The distance between them is about _____ inches on the map.

   That is about _____ miles.

Try This

4. The Statue of Liberty is number _____.

   The Sears Tower is number _____.

   The distance between them is about _____ inches on the map.

   That is about _____ miles.

5. Make up one of your own.

   _____ is number _____.

   _____ is number _____.

   The distance between them is about _____ inches on the map.

   That is about _____ miles.

96   ninety-six

Date **Time**

LESSON 4·9 Math Boxes

1. Write <, >, or =.

   5 × 5 _____ 4 × 4

   1 × 9 _____ 7 × 1

   6 × 2 _____ 2 × 8

   3 × 7 _____ 7 × 3

   SRB 13 56

2. Complete the Fact Triangle and write the fact family.

   ___ × ___ = ___

   ___ × ___ = ___

   ___ ÷ ___ = ___

   ___ ÷ ___ = ___

   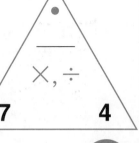

   SRB 55

3. Complete.

   Rule: ÷3

in	out
9	
15	
	7
	10

   SRB 52 203–204

4. Draw an array of 28 Xs arranged in 4 rows.

   How many Xs in each row? _____

   Write a number model for the array.

   _____

   SRB 64 65

5. Ages of 9 grandfathers:

   60, 54, 79, 80, 65, 74, 70, 65, 81

   mode = _____

   median = _____

   SRB 80 81

6. Fill in the empty frames.

   Rule: +1,000

   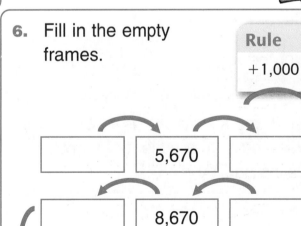

   SRB 200 201

97

LESSON 4·10 Coin-Toss Experiment

Work with a partner. You need 10 coins.

1. You will each toss all 10 coins 5 times.

   For each toss you make, record the number of HEADS and the number of TAILS in the table.

   Toss Record

Toss (10 coins)	HEADS	TAILS
1		
2		
3		
4		
5		
Total		

2. Use the information in both your partner's and your tables to fill in the blanks below.

   My total: HEADS _____   TAILS _____

   My partner's total: HEADS _____   TAILS _____

   Our partnership total: HEADS _____   TAILS _____

3. Record the number of HEADS and the number of TAILS for the whole class.

   Number of HEADS: _____   Number of TAILS: _____

4. Suppose a jar contains 1,000 pennies. The jar is turned over. The pennies are dumped onto a table and spread out. Write your best guess for the number of HEADS and TAILS.

   Number of HEADS: _____   Number of TAILS: _____

LESSON 4·10 Measuring Line Segments

Use your ruler to measure each line segment.

Measure to the nearest $\frac{1}{2}$ inch.

1. _____

   about _____ inches

2. _____

   about _____ inches

3. _____

   about _____ inches

Try This

Measure to the nearest $\frac{1}{4}$ inch.

4. _____

   about _____ inches

5. _____

   about _____ inches

Measure to the nearest $\frac{1}{8}$ inch.

6. _____

   about _____ inches

Date _____ Time _____

LESSON 4·10 Math Boxes

1. 18 marbles are shared equally.

   Each child gets 5 marbles.

   How many children are sharing?

   Use counters to solve.

   _____
   (unit)

   How many marbles are left over?

   _____
   (unit)

2. Is a 6-sided die more likely to land on an odd number or on an even number? Explain.

   _____

   _____

   _____

   _____

   _____

3. What is the perimeter of the rectangle?

   12 in. [rectangle] 22 in.

   The perimeter is _____.
   (unit)

4. Complete the Fact Triangle. Write the fact family.

   ___ × ___ = ___

   ___ × ___ = ___

   ___ ÷ ___ = ___

   ___ ÷ ___ = ___

5. Write the number that has

   5 hundreds
   7 thousands
   8 ones
   4 tens
   2 ten-thousands

   _____

   Read it to a partner.

6. Use >, <, or =.

   5,001 ____ 1,005

   55,001 ____ 51,005

   55,001 ____ 55,100

   505,105 ____ 505,105

100   one hundred

Date   Time

Lesson 4·11 Math Boxes

1. In the number 38,642

   the 4 means _____.

   the 8 means _____.

   the 6 means _____.

   the 3 means _____.

2. Put these numbers in order from smallest to largest.

   4,073   47,003   43,700   7,430

   _____ smallest

   _____

   _____

   _____ largest

3. Solve.   Unit

   12,469 + 10 = _____

   12,469 + 100 = _____

   12,469 + 1,000 = _____

   12,469 + 10,000 = _____

4. Which is more?

   $8.21 or $8.07 _____

   $0.07 or $0.48 _____

   $16.42 or $16.40 _____

5. Solve.   Unit

   ```
     6,000        400
       300          9
        20      8,000
   +     8    +    30
   ```

6. Write the number that is 100 more.

   76 _____

   300 _____

   471 _____

   8,634 _____

   5,925 _____

one hundred-one 101

LESSON 5·1 Place-Value Review

| Ten-Thousands | Thousands | Hundreds | Tens | Ones |

Follow the steps to find each number in Problems 1 and 2.

1. Write 6 in the ones place.
   Write 4 in the thousands place.
   Write 9 in the hundreds place.
   Write 0 in the tens place.
   Write 1 in the ten-thousands place.

   ___ ___ ___ ___ ___

2. Write 6 in the tens place.
   Write 4 in the ten-thousands place.
   Write 9 in the ones place.
   Write 0 in the hundreds place.
   Write 1 in the thousands place.

   ___ ___ ___ ___ ___

3. Compare the two numbers you wrote in Problems 1 and 2.

   Which is greater? _____

4. Complete.

   The 9 in 4,965 stands for 9 __*hundreds*__ or __*900*__.

   The 7 in 87,629 stands for 7 _____ or _____.

   The 4 in 48,215 stands for 4 _____ or _____.

   The 0 in 72,601 stands for 0 _____ or _____.

Continue the counts.

5. 4,707; 4,708; 4,709; _____; _____; _____

6. 7,697; 7,698; 7,699; _____; _____; _____

7. 903; 902; 901; _____; _____; _____

8. 6,004; 6,003; 6,002; _____; _____; _____

9. 47,265; 47,266; 47,267; _____; _____; _____

Lesson 5·1 Math Boxes

1. If a map scale shows that 1 inch represents 200 miles, then

2 inches represent _____ miles

3 inches represent _____ miles

5 inches represent _____ miles

1 inch
|——————|
0   200 miles

2. Put these numbers in order from smallest to largest.

54,752 _____

54,329 _____

54,999 _____

54,832 _____

3. Number of cookies in packages:

20, 24, 28, 30, 28, 26, 19, 24, 27

Put the data in order. Then find the median. Fill in the circle for the best answer.

Ⓐ 26 cookies   Ⓑ 25 cookies

Ⓒ 30 cookies   Ⓓ 1 cookie

4. Solve.

Unit

_____ = 80,000 − 40,000

_____ = 800,000 − 400,000

30,000 + 40,000 = _____

300,000 + 400,000 = _____

5. JANUARY

Su	M	Tu	W	Th	F	Sa
15	16	17	18	19	20	21
		31				

January 17th is a Tuesday. What is the date on the following Tuesday?

_____

6. Use your template. Trace two different polygons.

Lesson 5·2 Math Boxes

1. Use multiplication or division to complete these problems on your calculator.

Enter	Change to	How?
10	5	÷ 2
3	15	
6	60	
45	5	

SRB 52 53

2. Maximum number of treats: _____

Minimum number of treats: _____

Range of number of treats: _____

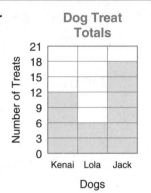

SRB 79

3. Find the total value. Fill in the circle for the best answer.

   4 $1
   3 Q
   6 D
   2 N
   7 P

   Ⓐ $4.36
   Ⓑ $5.17
   Ⓒ $4.67
   Ⓓ $5.52

4. Barry exercises every day. He walked 11 laps on both Monday and Thursday, 8 laps on Tuesday, and 9 laps on Wednesday. How many laps did he walk in all?

   _____
   (unit)

Total			
Part	Part	Part	Part
11	11	8	9

SRB 256 257

5. There are 3 cars. 4 people are riding in each car. How many people in all?

cars	people per car	people in all

Answer: _____
                 (unit)

Number model: _____

SRB 259 260

6. Complete.

A triangle has _____ sides.

A rectangle has _____ sides.

A square has _____ sides.

SRB 106–109

Lesson 5·3 Math Boxes

1. If a map scale shows that 1 cm represents 1,000 km, then

2 cm represent _____ km

9 cm represent _____ km

16 cm represent _____ km

0   1,000 km

2. Circle the largest number. Underline the smallest number.

946,487

946,800

946,793

946,200

3. Number of children in third grade classrooms:

31, 23, 21, 18, 28, 26, 22, 19, 30

What is the median?

_____ children

4. Solve.

Unit

70,000 + 80,000 = _____

700,000 + 800,000 = _____

_____ = 12,000 − 5,000

_____ = 120,000 − 50,000

5. JANUARY

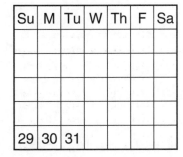

If January 31st is on a Tuesday, what day of the week will it be on February 1st?

_____

6. Trace all the quadrangles on your template.

105 one hundred five

LESSON 5·4 Working with Populations

Populations of 10 U.S. Cities

City	1990*	2000*
New York, NY	7,322,564	8,008,278
Chicago, IL	2,783,726	2,896,016
Houston, TX	1,630,553	1,953,631
Philadelphia, PA	1,585,577	1,517,550
Phoenix, AZ	983,403	1,321,045
Detroit, MI	1,027,974	951,270
Baltimore, MD	736,014	651,154
Nashville, TN	510,784	569,891
Sacramento, CA	369,365	407,018
Montgomery, AL	187,106	201,568

*U.S. Census Data

Use this table to solve the problems.

1. List the cities that lost population from 1990 to 2000.

2. Name a city where the population increased by more than 100,000.

3. Which city had the greatest change in total population? _____
   By about how many people did the population change? _____

4. In 1990, which city had a population about half that of Detroit, Michigan?

5. In 2000, which city had a population about double that of Montgomery, Alabama?

6. In 2000, which two cities combined had a population about the same as that of Chicago, Illinois?

106 one hundred six

Date _____ Time _____

LESSON 5·4 Math Boxes

1. Use multiplication or division to complete these problems on your calculator.

Enter	Change to	How?
6	24	× 4
24	3	
4	36	
36	6	

SRB 52 53

2. Here are the number of minutes 11 third graders spent doing homework: 25, 45, 55, 30, 35, 45, 60, 30, 45, 35, 40.

What is the mode? Fill in the circle with the best answer.

Ⓐ 45 minutes   Ⓑ 25 minutes

Ⓒ 60 minutes   Ⓓ 40 minutes

SRB 81

3. Complete.

20 dimes = $ _____

20 nickels = $ _____

20 quarters = $ _____

10 quarters and 7 dimes = $ _____

4. Travis jogged 20 minutes on Monday, 16 minutes on Tuesday, and 14 minutes on both Thursday and Friday. How many total minutes did he jog? _____
(unit)

Total			
Part	Part	Part	Part

SRB 256 257

5. 15 feet of ribbon. 3 feet in each yard of ribbon. How many yards of ribbon?

yards of ribbon	feet per yard	feet of ribbon in all

Answer: _____
Number model: _____
(unit)

6. Complete.

A pentagon has _____ sides.

A decagon has _____ sides.

An octagon has _____ sides.

one hundred seven **107**

Date _____ Time _____

LESSON 5·5 How Old Am I?

1. On what date were you born? _____

2. How old were you on your last birthday? _____ years old

3. About how many minutes old do you think you were on your last birthday? Mark an X next to your guess.

   _____ between 10,000 and 100,000 minutes

   _____ between 100,000 and 1,000,000 minutes

   _____ between 1,000,000 and 10,000,000 minutes

Use your calculator.

4. a. About how many days old were you on your last birthday? Do not include any leap year days. _____

   b. That's about how many hours? _____

   c. That's about how many minutes? _____

Try This

Adding Leap Year Days

5. a. List all of the leap years from the time you were born to your last birthday. _____

   b. That adds how many extra days to your last birthday? _____

   c. How many extra minutes? _____

6. Add the number of extra minutes to the number of minutes in your answer in Problem 4c. How many minutes are there in all? _____

7. On my last birthday, I was about _____ minutes old.

108 one hundred eight

Date _____ Time _____

LESSON 5·5 Math Boxes

1. Solve.

   Unit

   50,000 + 20,000 = _____

   500,000 + 200,000 = _____

   _____ = 90,000 − 50,000

   _____ = 900,000 − 500,000

2. Write the number.

   7 thousands
   8 tens
   5 ten-thousands
   1 one
   0 hundreds
   4 hundred-thousands

   ___ ___ ___ , ___ ___ ___

   SRB 18–21

3. Write the following amounts using a dollar sign and decimal point:

   2 dollar bills, 8 dimes, 9 pennies

   _____

   4 dimes, 6 pennies _____

   1 dollar bill, 4 pennies _____

   SRB 35

4. Solve using any method you wish.

   Unit

      907
   − 479
   ─────

   SRB 60–63

5. 13 crayons are shared equally among 3 children.

   How many crayons does each child get?

   _____ (unit)

   How many crayons are left over?

   _____ (unit)

   SRB 259 260

6. Use a straightedge. Draw a polygon with 5 sides.

   SRB 102–105

one hundred nine **109**

Date _____ Time _____

LESSON 5·6 Finding the Value of Base-10 Blocks

Exploration A:

Materials ☐ classroom supply of base-10 blocks

Work in a group.

1. Estimate the value of the base-10 blocks. Do not let anyone in your group see your estimate.

    Estimate: _____

2. Plan how your group will find the actual value of the blocks. Decide what each person will do to help. Carry out your plan.

3. What is the actual value of the base-10 blocks? _____

4. Write the estimates of your group and the actual value of the base-10 blocks in order from smallest to largest. Circle the actual value of the base-10 blocks.

    _____

    _____

5. Which estimate was closest to the actual value? _____

6. How many estimates were higher than the closest estimate? _____

7. How many estimates were lower than the closest estimate? _____

8. About how far was the highest estimate from the actual value? _____

9. About how far was the lowest estimate from the actual value? _____

10. How does your estimate compare to the actual value? _____

11. Describe how your group counted the blocks.

    _____

    _____

    _____

110   one hundred ten

Date   Time

LESSON 5·6 **Squares, Rectangles, and Triangles**

Exploration B:

Materials ☐ straightedge

A

H •   • E

D •   • B

G •   • F

•
C

Work on your own or with a partner.

1. Use your straightedge to draw line segments between points A and B, B and C, C and D, and D and A.

   What kind of shape did you draw? _____

2. Now draw line segments between points E and F, F and G, G and H, and H and E.

   What kind of shape did you draw? _____

3. Draw line segments between points E and G and between points F and H.

   How many different sizes of squares are there? _____

   How many squares in all? _____

4. How many different sizes of triangles are there? _____

   How many triangles in all? _____

5. How many rectangles are there that are not squares? _____

one hundred eleven   **111**

LESSON 5·6

Pattern-Block Perimeters

Exploration C:

Materials ☐ pattern blocks: square, large rhombus, small rhombus, triangle

Work on your own or with a partner.

1. Imagine each polygon is rolled along a line, starting at point *S*. Estimate the distance each polygon will roll after 1 full turn. Mark an X at the point you think the polygon will reach.

2. Check your estimate by rolling a pattern block that matches the polygon. Circle the point reached by the pattern block.

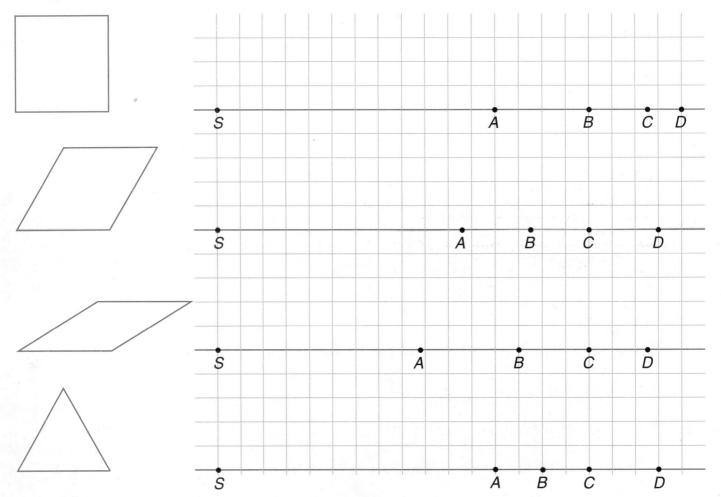

3. Which 3 shapes have about the same perimeter?

_____

4. Which of these 3 shapes has the largest area? _____

5. Which of the 4 shapes has the smallest area? _____

Date _____ Time _____

Lesson 5·6 Math Boxes

1. Circle the largest number. Underline the smallest number.

    56,689

    86,953

    90,865

    65,398

2. Write the following amounts in dollars-and-cents notation.

    5 dollar bills and 3 dimes _____

    4 dollar bills and 6 dimes _____

    1 dollar bill and 1 penny _____

3. The population of Chandler, Arizona, nearly doubled from 1990 to 2000. If 89,862 people lived in Chandler in 1990*, estimate about how many people lived there in 2000.

    Ballpark estimate:

    _____

    Answer: _____
    (unit)

    *U.S. Census Bureau

4. Complete.

fishbowls	fish per bowl	fish in all
4	4	?

    Answer: _____
    (unit)

    Number model: _____

5. Complete.

    _____ days in a week

    _____ days in two weeks

    _____ days in three weeks

    _____ days in four weeks

6. Measure each side of the quadrangle to the nearest half-centimeter.

    _____ cm

    _____ cm   _____ cm

    _____ cm

    Another name for this quadrangle is a

    _____ .

one hundred thirteen **113**

| Date | Time |

LESSON 5·7 Place Value in Decimals

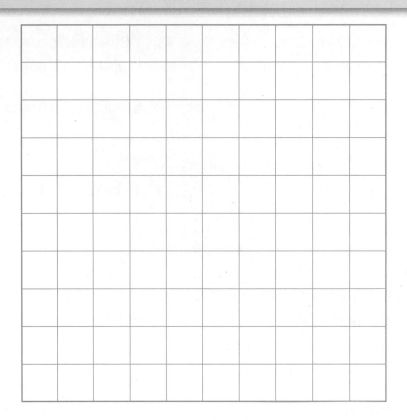

If the grid is ONE, then which part of each grid is shaded?

Write a fraction and a decimal below each grid.

1.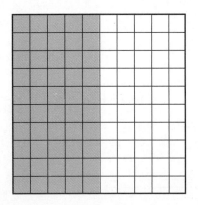

2.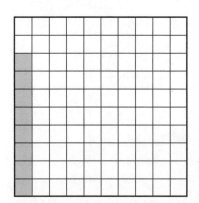

3.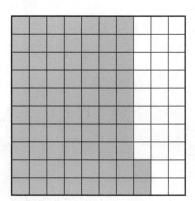

fraction: _____   fraction: _____   fraction: _____

decimal: _____   decimal: _____   decimal: _____

LESSON 5·7 Place Value in Decimals continued

4. Which decimal in each pair is greater? Use the grids in Exercises 1–3 to help you.

   0.5 or 0.08 _____   0.08 or 0.72 _____   0.5 or 0.72 _____

Color part of each grid to show the decimal named.

5. Color 0.7 of the grid.   6. Color 0.07 of the grid.   7. Color 0.46 of the grid.

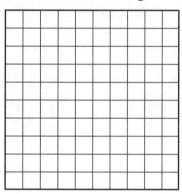 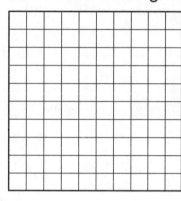 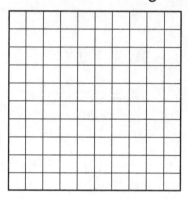

8. Write 0.7, 0.07, and 0.46 in order from smallest to largest.

   Use the grids in Exercises 5–7 to help you. _____ _____ _____

Try This

Color part of each grid to show the fraction named.

9. Color $\frac{4}{10}$ of the grid.   10. Color $\frac{1}{2}$ of the grid.   11. Color $\frac{23}{100}$ of the grid.

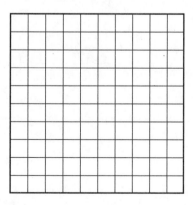 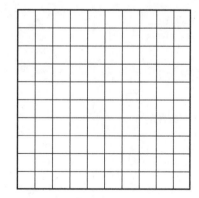 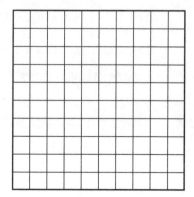

12. Write $\frac{23}{100}$ as a decimal. _____

one hundred fifteen 115

Date _____ Time _____

LESSON 5·7 Math Boxes

1. Solve.

   Unit

   16 + 9 = _____

   16 + 90 = _____

   16 + 900 = _____

   16 + 9,000 = _____

   16 + 90,000 = _____

2. For the number 5,749,862

   the 4 means __40,000__

   the 5 means _____

   the 8 means _____

   the 7 means _____

   the 9 means _____

   SRB 18–21

3. What is another name for 3 dollar bills and 6 pennies? Fill in the circle for the best answer.

   Ⓐ $3.60

   Ⓑ $3.06

   Ⓒ $30.60

   Ⓓ 3.06¢

   SRB 35

4. Solve using any method you wish.

   Unit

   199
   +499
   ─────

   SRB 60–63

5. Write a division story by filling in the blanks. There are 48 _____ in 6 rows. How many _____ are in each row? _____

   Write a number model.

   _____

   SRB 250–253 260

6. Use a straightedge. Draw a 7-sided polygon.

   SRB 102–105

116   one hundred sixteen

Lesson 5·8 Exploring Decimals

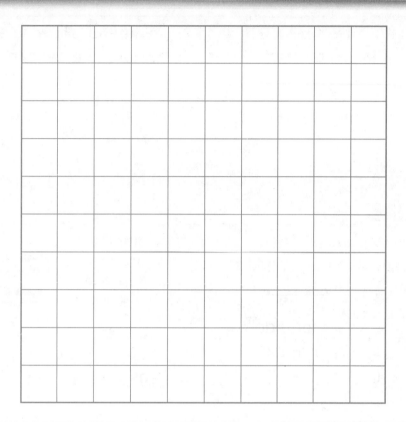

A	B	C	D
__13__ hundredths	__1__ tenth, __3__ hundredths	0. __13__	__13/100__
_____ hundredths	_____ tenths, _____ hundredths	0. _____	
_____ hundredths	_____ tenths, _____ hundredths	0. _____	
_____ hundredths	_____ tenths, _____ hundredths	0. _____	
_____ hundredths	_____ tenths, _____ hundredths	0. _____	
_____ hundredths	_____ tenths, _____ hundredths	0. _____	
_____ hundredths	_____ tenths, _____ hundredths	0. _____	

LESSON 5·8 Math Boxes

1. Which number is the smallest? Fill in the circle for the best answer.

- ○ **A.** 693,971
- ○ **B.** 809,178
- ○ **C.** 97,987
- ○ **D.** 488,821

2. Write the following amounts in dollars-and-cents notation.

2 dollar bills and 3 dimes _____

7 dimes and 9 pennies _____

5 pennies _____

3. The land area of Alaska is 571,951 square miles. The land area of Texas is 261,797 square miles.* Estimate about how much bigger Alaska is than Texas.

Ballpark estimate:

_____

Answer: _____ (unit)

*data from *The World Almanac and Book of Facts 2004*

4. Complete.

tricycles	wheels per tricycle	wheels in all
?	3	24

Answer: _____ (unit)

Number model:

_____

5. Complete.

1 hour = _____ minutes

1 day = _____ hours

1 week = _____ days

1 year = _____ months

6. Measure each side of the polygon to the nearest half-inch.

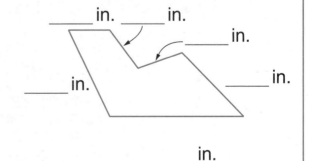

_____ in.   _____ in.

_____ in.

_____ in.   _____ in.

_____ in.

A 6-sided polygon is called a _____.

118 one hundred eighteen

Date _____ Time _____

LESSON 5·9 Decimals for Metric Measurements

1. Fill in the missing information. Put longs and cubes end to end on a meterstick to help you.

Length in Centimeters	Number of Longs	Number of Cubes	Length in Meters
24 cm	2	4	0.24 m
36 cm	___	___	___ m
___ cm	0	3	___ m
8 cm	___	___	___ m
___ cm	___	___	0.3 m
___ cm	4	3	___ m

Work with a partner. Each partner uses base-10 blocks to make one length in each pair. Compare the lengths and circle the one that is greater.

2. 0.09 or 0.12   3. 0.24 or 0.42   4. 0.10 or 0.02

5. 0.18 or 0.5   6. 0.2 or 0.08   7. 0.3 or 0.24

Follow these directions on the ruler below. Use base-10 blocks and a meterstick to help you.

8. Make a dot at 4 cm and label it with the letter A.

9. Make a dot at 0.1 m and label it with the letter B.

10. Make a dot at 0.15 m and label it with the letter C.

11. Make a dot at 0.08 m and label it with the letter D.

one hundred nineteen

Date _____ Time _____

Lesson 5·9 Math Boxes

1. Write the number that has
   6 in the ones place
   4 in the tenths place
   3 in the hundredths place
   2 in the thousandths place

   ___.___ ___ ___

2. If each grid is ONE, what part of each grid is shaded? Write the decimal.

   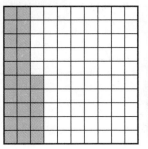 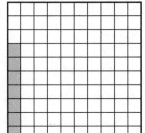

   _____   _____

3. Use addition and subtraction to complete these problems on your calculator.

Enter	Change to	How?
894	12,894	
1,366	966	
627,581	628,581	
43,775	43,175	

4. Draw a 3 × 7 rectangle.

   Number model: ___ × ___ = ___

   Area: ___ square units

5. For the number 4,963,521

   4 means 4,000,000

   3 means _____

   1 means _____

   6 means _____

   9 means _____

6. Draw an example of a cylinder.

120 one hundred twenty

LESSON 5·10   How Wet? How Dry?

1. Use the scale at the left and the map on page 221 of the *Student Reference Book*. Make a dot for the level of precipitation in each of the following cities: Seattle, Omaha, Birmingham, and Tampa. Write the name of the city next to the dot.

2. Which city gets about 5 centimeters less rain than Mobile?

   _____

3. Which city gets about half as much rain as Omaha?

   _____

4. Which city gets about 4 times as much rain as Seattle?

   _____

5. A decimeter is 10 centimeters. Which cities on the map get at least 1 decimeter of rain?

   _____

(Scale at left: 0–22 cm, with Mobile marked at about 15 cm)

Did You Know?

According to the National Geographic Society, the rainiest place in the world is Mount Waialeale in Hawaii. It rains an average of about 1,170 centimeters a year on Mount Waialeale.

Try This

6. Suppose it rained 1,170 centimeters in your classroom. Would the water reach the ceiling?

   _____ millimeters = 1,170 centimeters = _____ meters

   Answer: _____

one hundred twenty-one   **121**

Date _____ Time _____

LESSON 5·10 Math Boxes

1. Color 0.6 of the grid.

   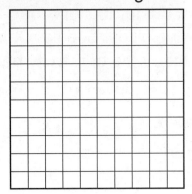

2. Complete the bar graph.

   Lily ran 4 miles.

   Meg ran 3 miles.

   Rita ran 6 miles.

   Median miles run: _____

3. Write <, >, or =.

   0.65 ____ 0.56

   0.07 ____ 0.7

   0.098 ____ 0.102

   73.4 ____ 75.2

4. What is the perimeter of the trapezoid?

   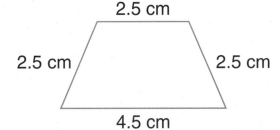

   ○ A. 10 cm   ○ B. 11 cm
   ○ C. 12 cm   ○ D. 13 cm

5. Write the number that has
   2 in the ones place
   6 in the tenths place
   7 in the hundredths place

   ___ . ___ ___

6. This polygon has ___ sides.

   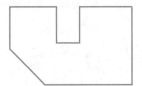

   It is called a _____.

122 one hundred twenty-two

LESSON 5·11 More Decimals

Use your Place-Value Book to help you. Write the number that matches each description.

1. 0 in the ones place
   8 in the tenths place
   _____

2. 1 in the ones place
   3 in the tenths place
   _____

3. 2 in the ones place
   7 in the hundredths place
   0 in the tenths place
   _____

4. 0 in the hundredths place
   6 in the ones place
   8 in the thousandths place
   0 in the tenths place
   _____

5. Read each of the decimals in Problems 1–4 to a partner.

Write each number below as a decimal.

6. nine-tenths _____

7. thirty-thousandths _____

8. fifty-three hundredths _____

9. sixty and four-tenths _____

10. seven and seven-thousandths
    _____

11. sixty and four-hundredths
    _____

Unit: meter

Fill in the missing numbers.

12.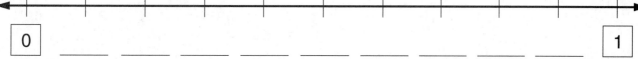

13.

one hundred twenty-three **123**

Date _____ Time _____

LESSON 5·11 Math Boxes

1. Write the number that has

   5 in the tenths place
   4 in the hundredths place
   1 in the ones place
   6 in the thousandths place

   ___.___ ___ ___

2. How much of this grid is shaded? Fill in the circle for the best answer.

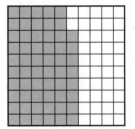

   ○ A. 0.58
   ○ B. 58
   ○ C. 5.8
   ○ D. 0.6

3. Use addition and subtraction to complete these problems on your calculator.

Enter	Change to	How?
629	18,629	
2,411	411	
456,972	450,972	
28,684	28,084	

4. Draw a shape with an area of 16 square units.

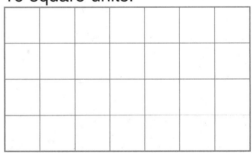

How many sides does your shape have?

_____ sides

5. For the number 3,975,081

   5 means __5,000__

   1 means _____

   7 means _____

   9 means _____

   3 means _____

6. An example of a sphere is a ball. Draw or name another sphere.

124 one hundred twenty-four

Date | Time

LESSON 5·12 Length-of-Day

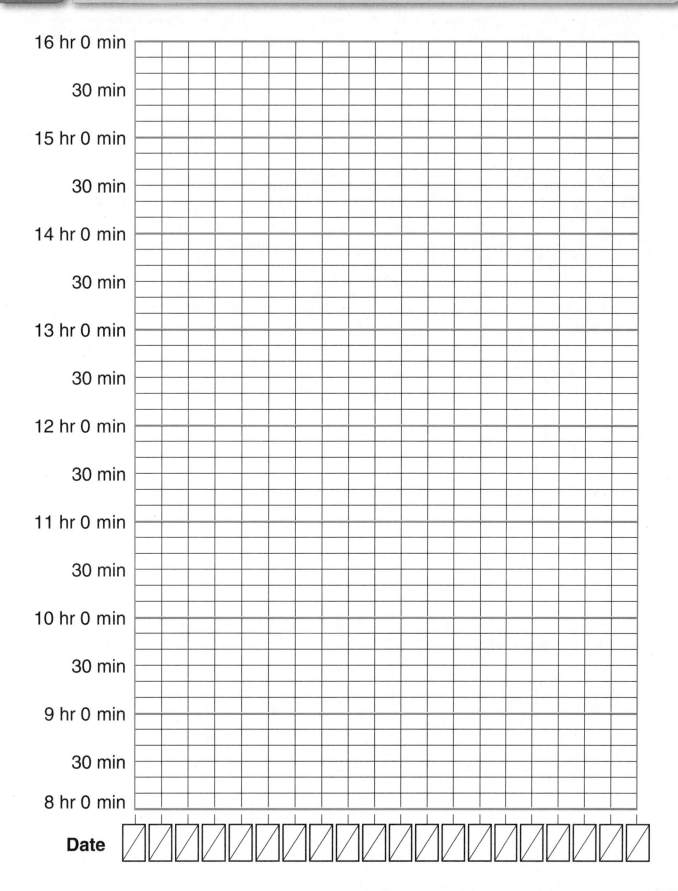

Date _____ Time _____

LESSON 5·12 Math Boxes

1. Color 0.08 of the grid.

   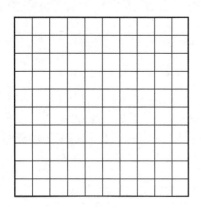

2. What is the maximum number of points? _____

   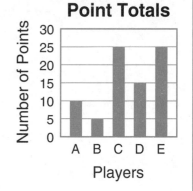

   What is the mode? _____

3. Which is more?

   1.36 or 1.6 _____

   0.4 or 0.372 _____

   0.69 or 0.6 _____

   0.7 or 0.09 _____

4. Find the perimeter of the octagon.

   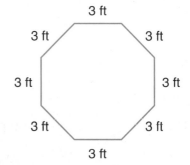

   Perimeter = _____ (unit)

5. Write the number that has
   4 in the tenths place
   0 in the hundredths place
   6 in the ones place
   9 in the thousandths place

   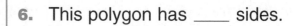 .

6. This polygon has ____ sides.

   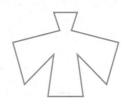

   Name the shape. _____

126   one hundred twenty-six

Date　　　　　　　　　　　　Time

LESSON 5·13　Math Boxes

1. Draw a line that will divide the rectangle into 2 equal parts.

2. Complete.

   A triangle has _____ sides and _____ angles.

   A quadrangle has _____ sides and _____ angles.

   SRB 106–109

3. Draw a quadrangle.

   SRB 108 109

4. Draw a polygon.

   SRB 102 103

5. Use your template. Draw a rhombus and a square. How are they alike?

   _____
   _____

6. Circle the pictures that show 3-dimensional shapes.

   SRB 112–115

one hundred twenty-seven　**127**

LESSON 6·1 Line Segments, Rays, and Lines

1. Write S next to each line segment. Write R next to each ray. Write L next to each line.

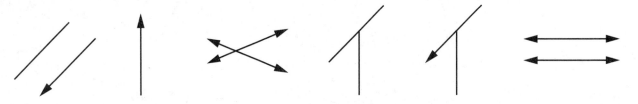

Points D, T, Q, and M are marked. Use a straightedge to draw the following.

2. Draw $\overline{QT}$. Draw $\overrightarrow{DT}$. Draw $\overleftrightarrow{MQ}$.

D •   T •

M •   Q •

Draw a line segment between each pair of points. How many line segments did you draw?

Example:

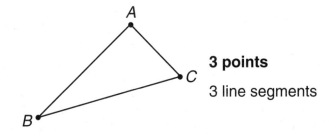

3 points
3 line segments

3.   • P

  A •

  • L

  • U

4 points

_____ line segments

4.   • R

  O •
              • E

  S •
              • I

5 points

_____ line segments

128 one hundred twenty-eight

Date   Time

LESSON 6·1 Math Boxes

1. Complete the number-grid puzzle.

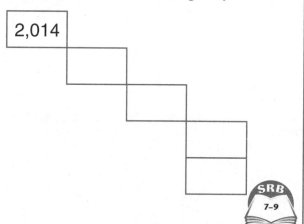

2. If the grid is ONE, then what part of the grid is shaded? Write a fraction and a decimal.

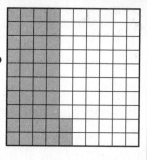

_____ = _____
(fraction)   (decimal)

3. What is a fair way to decide who should go first in a game?

_____
_____
_____

4. Cross out the names that do not belong in this name-collection box.

0.1	.01	$\frac{10}{100}$
$\frac{1}{100}$	one-hundredth	
	0.10	10
$\frac{1}{10}$	one-tenth	

5. In the number 2.673,

the 6 means __6 tenths__.

the 3 means _____.

the 7 means _____.

the 2 means _____.

6. Complete the Fact Triangle. Write the fact family.

72, ×, ÷, 8

_____
_____
_____
_____

one hundred twenty-nine   **129**

LESSON 6·2
Geometry Hunt

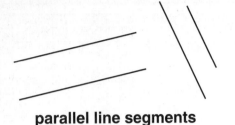

parallel line segments

intersecting line segments

Part 1 (Use with Lesson 6-2.)

Look for things in the classroom or hallway that are parallel. Look for things that intersect. List these things below or draw a few of each of them on another sheet of paper.

Parallel

_____

_____

_____

Intersecting

_____

_____

_____

Part 2 (Use with Lesson 6-3.)

Look for things in the classroom or hallway that have one or more right angles. List these things below or draw a few of them on another sheet of paper.

_____

_____

_____

130 one hundred thirty

Date **Time**

LESSON 6·2 Math Boxes

1. The grid is ONE.

   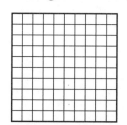

   Shade 0.06 of   Shade 0.25 of
   the grid.   the grid.

   Write the larger number.

   _____

2. 9 boxes of muffins. 6 muffins per box. How many muffins in all?

   _____
   (unit)

   Write a number model:

   _____ × _____ = _____

3. Use some or all of the cards to write different names for the target number.

    3   2   5   4   6   12

   target number

   _____

   _____

   _____

   _____

4. **Favorite Vegetables**

   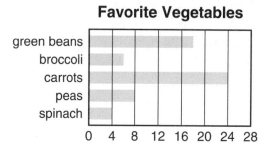

   Which vegetable is the least favorite? _____

5. Draw a ray, $\overrightarrow{AB}$. Draw a line segment, $\overline{CD}$. Draw a line, $\overleftrightarrow{EF}$.

   • A   • B

   • C   • D

   • E   • F

6. Solve.

   Double 2 _____

   Double 10 _____

   Double 75 _____

   Double 1,000 _____

   Double 1,500 _____

one hundred thirty-one **131**

LESSON 6·3 **Turns**		

Use your connected straws to show each turn.
Draw a picture of what you did.
Draw a curved arrow to show the direction of the turn.

Example:

right $\frac{1}{4}$ turn (clockwise)	**1.** right $\frac{1}{2}$ turn (clockwise)	**2.** left $\frac{1}{4}$ turn (counterclockwise)
3. left $\frac{3}{4}$ turn (counterclockwise)	**4.** right $\frac{3}{4}$ turn (clockwise)	**5.** left $\frac{1}{2}$ turn (counterclockwise)

Date _____ Time _____

LESSON 6·3 Math Boxes

1. Complete the number-grid puzzle.

   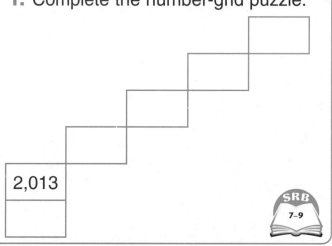

   2,013

2. If the grid is ONE, then what part of the grid is shaded? Circle the best answer.

   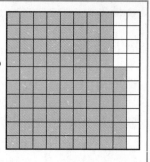

   A   0.86   B   0.68

   C   08.6   D   86

3. Lisa tosses a coin. How likely is the coin to land on HEADS? Circle one:

   More likely to land on HEADS than on TAILS

   Equally likely to land on HEADS or on TAILS

   Less likely to land on HEADS than on TAILS

4. Cross out the names that do not belong in the name-collection box.

.05	$\frac{5}{10}$	5.0
$\frac{5}{100}$	five-tenths	0.05
	five-hundredths	0.50

5. In the number 34.972,

   the 9 means ___9 tenths___.

   the 7 means _____.

   the 3 means _____.

   the 4 means _____.

   the 2 means _____.

6. Complete the Fact Triangle. Write the fact family.

   _____

   _____

   _____

   _____

one hundred thirty-three   133

Date _____ Time _____

LESSON 6·4 **Exploring Triangles**

Part 1

Follow these steps:

1. Find the three points on the right.

2. Use a straightedge to connect each pair of points with a line segment.

3. What figure have you drawn?

   _____

B •

• C

• A

Part 2

Write all six 3-letter names that are possible for your triangle. The first letter of each name is given below.

A_____   A_____   B_____   B_____   C_____   C_____

Part 3

Work with a group.

Make triangles with straws and twist-ties. Make at least one of each of the following kinds of triangles.

◆ all 3 sides the same length

◆ only 2 sides the same length

◆ no sides the same length

◆ 1 angle larger than a right angle

◆ all 3 angles smaller than a right angle

Part 4

Measure each side of the triangle you drew in Part 1 to the nearest $\frac{1}{4}$ inch.

side AB _____ in.   side BC _____ in.   side CA _____ in.

134   one hundred thirty-four

Date Time

LESSON 6·4 Math Boxes

1.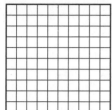

   Shade 0.6 of the grid.   Shade 0.30.

   Write the larger number. _____

2. 13 crayons are shared equally among 3 children. How many crayons does each child get?

   Ⓐ 3 crayons, 4 left over

   Ⓑ 4 crayons, 1 left over

   Ⓒ 3 crayons, 1 left over

   Ⓓ 4 crayons, 2 left over

3. Fill in the name-collection box with at least five equivalent names.

4. At what ages can the most girls arm hang for 8 seconds?

   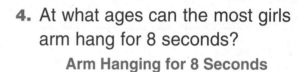

   _____

5. Draw $\overrightarrow{AT}$. Draw $\overline{BY}$. Draw $\overleftrightarrow{ME}$.

   • A   • T

   • B   • Y

   • M   • E

6. Complete.

in	out
16	
	240
225	
133	
	1,000

   Rule: double

one hundred thirty-five **135**

LESSON 6·5

Exploring Quadrangles

Part 1

Use a straightedge. Connect points to form a quadrangle.

B •

A •

D •   • C

Part 2

Write all 4-letter names that are possible for your quadrangle. The first letter of each name is given below.

A_____   A_____   B_____   B_____
C_____   C_____   D_____   D_____

Part 3

Work in a group.

Make quadrangles with straws and twist-ties. Make at least one of each of the following kinds of quadrangles.

◆ all 4 sides equal in length
◆ 2 pairs of equal-length sides, but opposite sides not equal in length
◆ 2 pairs of equal-length opposite sides
◆ only 2 parallel opposite sides
◆ only 1 pair of equal-length opposite sides

Part 4

Measure each side of the quadrangle you drew in Part 1 to the nearest $\frac{1}{2}$ centimeter.

side AB _____ cm   side BC _____ cm   side CD _____ cm   side DA _____ cm

Try This

The perimeter of my quadrangle is about _____ centimeters.

Date   Time

Lesson 6·5 Math Boxes

1. The grid is ONE. Shade 0.41 of the grid.

   Write the fraction that shows how much is shaded.

   0.41 = _____

2. Circle the pair of lines that are parallel.

3. Fill in the oval for the best answer. The turn of the angle is

   ⊙ less than a $\frac{1}{2}$ turn.
   ⊙ less than a $\frac{1}{4}$ turn.
   ⊙ greater than a $\frac{1}{2}$ turn.
   ⊙ a full turn.

4. Draw a ray, $\overrightarrow{DO}$. Draw a line segment, $\overline{RE}$. Draw a line, $\overleftrightarrow{MI}$.

5. Draw a shape with 4 sides that are all equal in length.

   This shape is a _____.

6. Complete.

   ×4   ÷2

   6   ( )   12   ( )

   ( )   48   ( )   ( )

LESSON 6·6 Exploring Polygons

Part 1

1. Use a straightedge and draw $\overline{AB}$, $\overline{BC}$, $\overline{CD}$, $\overline{DE}$, and $\overline{EA}$.

2. What kind of polygon did you draw?

   _____

3. Write 4 or more possible letter names for the polygon.

   _____   _____   _____

   _____   _____   _____

Part 2

Work in a group to make polygons with straws and twist-ties. Your teacher will tell you how many sides each polygon should have.

Make at least one of each of the following kinds of polygons.

◆ all sides equal in length, and all angles equal in size (the amount of turn between sides)

◆ all sides equal in length but not all angles equal in size

◆ any polygon having the assigned number of sides

LESSON 6·6

Exploring Polygons continued

Part 3

A **regular polygon** is a polygon in which all the sides are equal and all the angles are equal.

Below, trace the smaller of each kind of *regular* polygon from your Pattern-Block Template.	Below, trace all the polygons from your Pattern-Block Template that are *not* regular polygons.

Part 4

Measure each side of the polygon you drew in Part 1 to the nearest $\frac{1}{2}$ centimeter.

side AB about _____ cm

side BC about _____ cm

side CD about _____ cm

side DE about _____ cm

side EA about _____ cm

Try This

The perimeter of my polygon is about _____ cm.

one hundred thirty-nine **139**

Date _____ Time _____

LESSON 6·6 Math Boxes

1. A pentagon has

_____ sides,

_____ vertices,

and _____ angles.

Draw a pentagon.

2. If each grid is ONE, what part of each grid is shaded? Write the decimal.

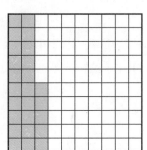 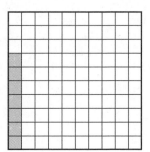

_____   _____

Circle the larger number.

3. Write <, >, or =.

0.45 _____ 0.54

1.07 _____ 1.7

2.3 _____ 0.23

10.8 _____ 10.80

0.2 _____ 2.0

4. 64 slices of pizza. 8 people. How many slices per person?

Fill in the oval for the best answer.

◯ 64 + 8
◯ 64 − 8
◯ 64 × 8
◯ 64 ÷ 8

5. Draw line segments to form a quadrangle.

• M   • A

• H
    • T

Which letter names the right angle?

_____

6. Complete the Fact Triangle. Write the fact family.

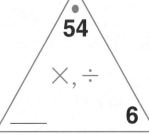

_____

_____

_____

_____

140   one hundred forty

Date	Time

LESSON 6·7 Drawing Angles

Draw each angle as directed by your teacher.
Record the direction of each turn with a curved arrow.

Part 1

A •   B •

                    C •

Part 2

                    S •

    •
    R

        T •

one hundred forty-one   **141**

Date _____ Time _____

LESSON 6·7 Math Boxes

1. The grid is ONE. Shade $\frac{57}{100}$ of the grid.

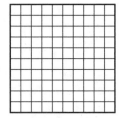

Write the decimal that tells how much of the grid is shaded.

_____

2. Circle the pairs of lines that intersect.

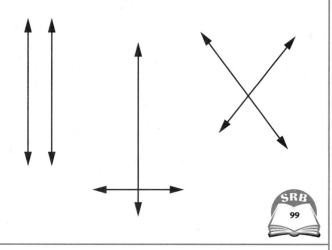

3. Draw an angle that is less than a $\frac{1}{4}$ turn.

4. Draw a ray, $\overrightarrow{SO}$. Draw a line segment, $\overline{LA}$. Draw a line, $\overleftrightarrow{TI}$.

· ·

· ·

· ·

5. Circle the regular polygons.

6. Complete.

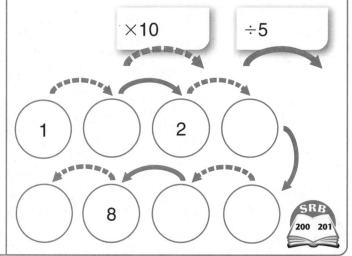

142 one hundred forty-two

Date	Time

LESSON 6·8 Marking Angle Measures

Connect 2 straws with a twist-tie. Bend the twist-tie at the connection to form a vertex.

♦ Place the straws with the vertex on the center of the circle.

♦ Place both straws pointing to 0°.

Keep one straw pointing to 0°. Move the other straw to form angles.

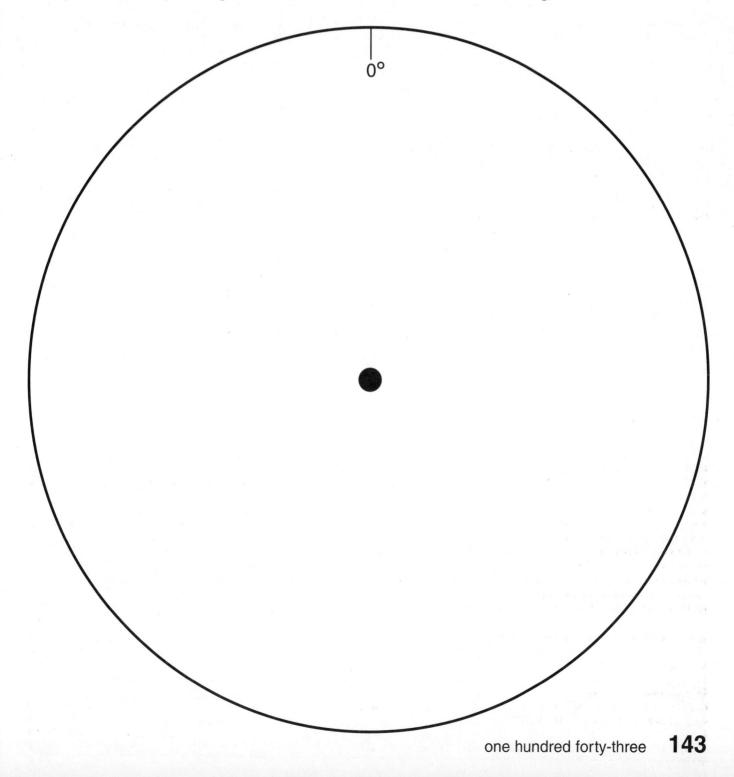

one hundred forty-three **143**

Date   Time

LESSON 6·8   Measuring Angles

Use your angle measurer to measure the angles on this page.
Record your measurements in the table. Then circle the right angle below.

Angle	Measurement
A	about _____°
B	about _____°
C	between _____° and _____°
D	about _____°
E	about _____°
F	about _____°

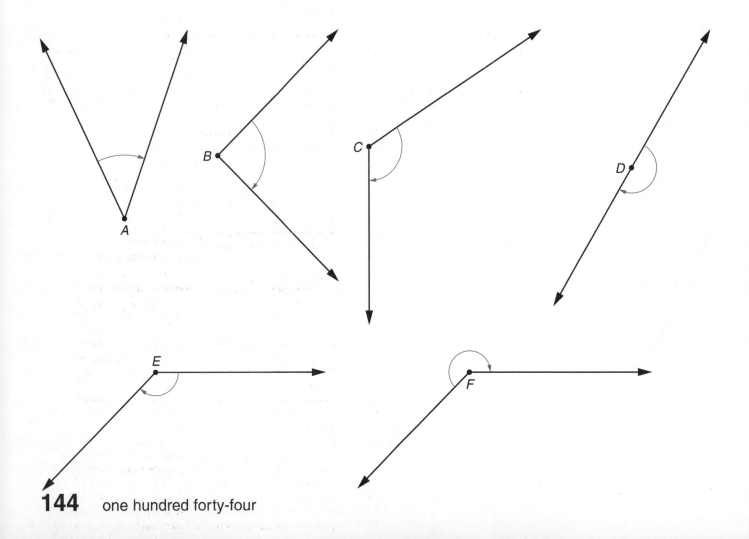

144   one hundred forty-four

LESSON 6·8 Math Boxes

1. Continue the pattern.

   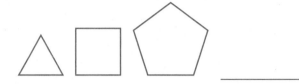

2. If each grid is ONE, what part of each grid is shaded? Write the decimal.

   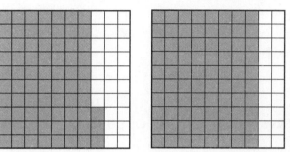

   _____   _____

   Circle the smaller number.

3. Write these numbers in order from smallest to largest:
   0.2; 0.02; 0.19

   _____ _____ _____
   smallest   largest

4. Write a number model that matches the diagram.

vans	people per van	people in all
7	7	49

   Number model: _____

5. Write the letter that names the right angle.

   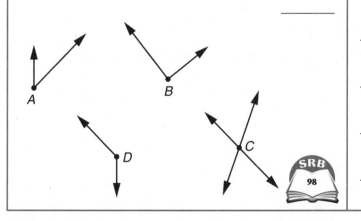

   _____

6. Complete the Fact Triangle. Write the fact family.

   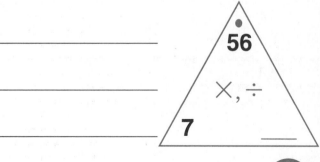

   _____

   _____

   _____

   _____

one hundred forty-five 145

LESSON 6·9 Symmetric Shapes

Each picture shows one-half of a letter. The dashed line is the line of symmetry. Guess what the letter is. Then draw the other half of the letter.

1.
2.
3.
4.

Draw the other half of each symmetric shape below.

5.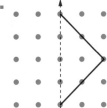
6.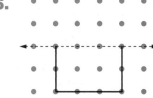

7.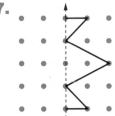
8.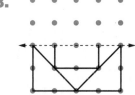

9. The picture at the right shows one-fourth of a symmetric shape, and two lines of symmetry. Draw the mirror image for each line of symmetry.

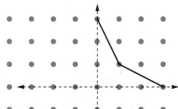

Try This

10. The finished figure in Problem 9 has 2 more lines of symmetry. Draw them.

146 one hundred forty-six

Date   Time

LESSON 6·9   Math Boxes

1. 3 people share 14 pennies.

   Each person gets _____ pennies.

   There are _____ pennies left.

2. A baker packed 8 boxes of cupcakes. She packed 4 chocolate and 4 white cupcakes in each box. How many cupcakes did she pack in all?

   _____
   (unit)

3. Draw a quadrangle with exactly one right angle. Label the vertices A, B, C, D. Which letter names the right angle?

   Angle ____

4. Use your template. Draw a shape that has 6 vertices.

   This shape is a _____

5. Describe the angle.

   Fill in the circle for the best answer.

   ○ A. greater than a $\frac{1}{4}$ turn

   ○ B. less than a $\frac{1}{4}$ turn

   ○ C. greater than a $\frac{1}{2}$ turn

   ○ D. one full turn

6. Estimate. A package of cookies costs $2.09. About how much do 3 packages cost? Show the number model for your estimate.

   About _____

   Number model:

   _____

one hundred forty-seven **147**

LESSON 6·10 Base-10 Block Decimal Designs

Exploration C:

Materials
- ☐ base-10 blocks (cubes, longs, and flats)
- ☐ 10-by-10 grids (*Math Journal 1*, p. 149)
- ☐ crayons or colored pencils

Think of the flat as a unit, or ONE. Remind yourself of the answers to the following questions:

◆ How many cubes would you need to cover the whole flat?

◆ How much of the flat is covered by 1 cube? By 1 long?

Follow these steps:

Step 1 Make a design by putting some cubes on a flat.

Step 2 Copy your design in color onto one of the grids on journal page 149.

Step 3 How much of the flat is covered by the cubes in your design? To help you find out, exchange as many cubes as you can for longs.

Step 4 Figure out which decimal tells how much of the flat is covered by cubes. Write the decimal under the grid that has your design on it.

Repeat steps 1–4 to create and count other designs.

Example:

Step 1: Make a design on a flat with cubes.

Step 2: Copy the design onto a grid.

Step 3: Exchange cubes for longs. Figure out how much of the flat is covered.

Step 4: Write the decimal under the grid.

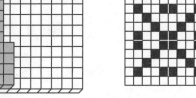

Step 4
Decimal: 0.24

148 one hundred forty-eight

Date _____ Time _____

LESSON 6·10   10 × 10 Grids

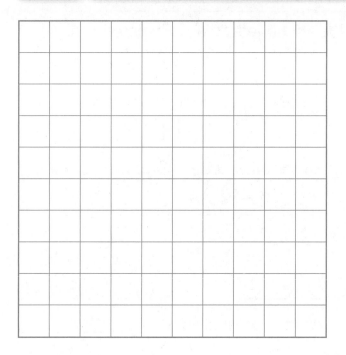

Decimal: _____

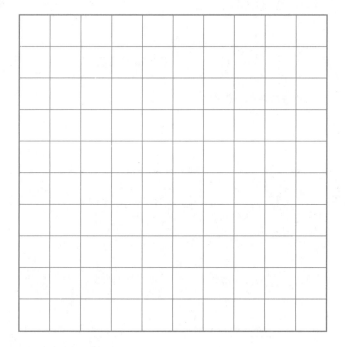

Decimal: _____

Decimal: _____

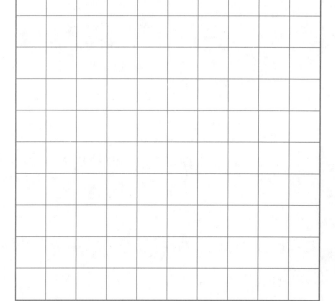

Decimal: _____

LESSON 6·10 Math Boxes

1. Draw a line segment, $\overline{DI}$, parallel to the line, $\overleftrightarrow{PO}$. Draw a ray, $\overrightarrow{LA}$, that intersects the line, $\overleftrightarrow{TW}$.

2. These letters are *Symmets*:

H, T, M, A

These letters are not *Symmets*:

F, J, R, S

Write other letters that are *Symmets*:

_____

_____

3. What is the difference in points between Players B and C?

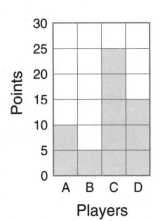

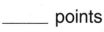 points

What are the total points scored for all players?

_____ points

4. Write the numerals.

forty-hundredths _____

four-tenths _____

six-tenths _____

sixteen-hundredths

_____

5. Connect 4 points. Label the points.

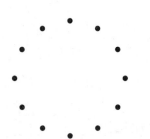

What shape did you draw?

_____

6. Multiply.

2 × 5 = _____

7 × 3 = _____

_____ = 5 × 5

_____ = 2 × 7

_____ = 4 × 6

| Date | Time |

LESSON 6·11 Symmetry

If a shape can be folded in half so that the two halves match exactly, the shape is **symmetric.** We also say that the shape has **symmetry.**

The fold line is called the **line of symmetry.** Some symmetric shapes have just one line of symmetry. Others have more.

1 line of symmetry

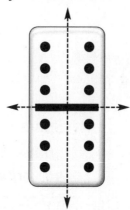
2 lines of symmetry

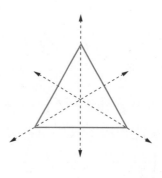
3 lines of symmetry

1. Which of the following shapes is **not** symmetric? _____

   a.

   b.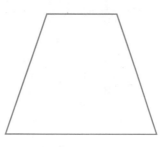

   c.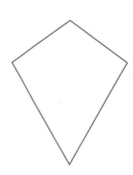

   d.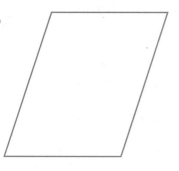

   e.

   f.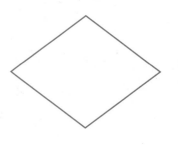

2. Draw all the lines of symmetry on the shapes that are symmetric.

one hundred fifty-one **151**

Date _____ Time _____

LESSON 6·11 Math Boxes

1. 4 people share 18 crayons.

   Each person gets _____ crayons.

   There are _____ crayons left.

2. Dale had 9 toy cars. Jim had 4 less than twice as many as Dale. How many toy cars did Jim have?

   _____
   (unit)

3. Circle the right triangles. Use the corner of a piece of paper to check.

   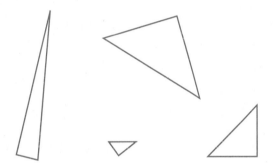

4. Solve the riddle.
   I have four sides. My opposite sides are equal in length. One pair of my sides is longer than the other pair. Draw my shape.

   I am a _____.

5. Fill in the circle for the best answer. The turn of the angle is

   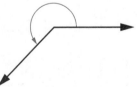

   ○ A. greater than a $\frac{3}{4}$ turn.

   ○ B. less than a $\frac{1}{4}$ turn.

   ○ C. greater than a $\frac{1}{2}$ turn.

   ○ D. a full turn.

6. Estimate. 1 bag of marbles costs $1.45. About how much do 2 bags cost? Show the number model for your estimate.

   About _____

   Number model:

   _____

one hundred fifty-two

LESSON 6·12 Pattern-Block Prisms

Work in a group.

1. Each person chooses a different pattern-block shape.

2. Each person then stacks 3 or 4 of the shapes together to make a prism. Use small pieces of tape to hold the blocks together.

   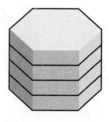

3. Below, carefully trace around each face of your prism. Then trace around each face of 2 or 3 more prisms on a separate sheet of paper. Ask someone in your group for help if you need it. Share prisms with other people in your group.

one hundred fifty-three **153**

Date _____ Time _____

LESSON 6·12 Math Boxes

1. Draw a line, $\overleftrightarrow{AB}$, parallel to line segment, $\overline{CD}$. Draw a ray, $\overrightarrow{EF}$, that intersects the ray, $\overrightarrow{GH}$.

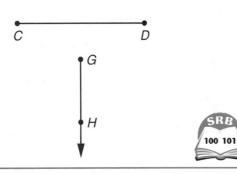

2. Draw all the lines of symmetry.

There are _____ lines of symmetry.

3. Number of days for one revolution around the sun:

Mercury	88
Venus	225
Earth	365
Mars	687

Which planet takes the fewest days to revolve around the sun?

Fill in the circle for the best answer.

○ A. Mercury   ○ C. Venus
○ B. Earth   ○ D. Mars

4. Write the numerals.

five-tenths _____

five-hundredths _____

three-tenths _____

three-hundredths _____

5. Connect 3 points to make a right triangle. Label the points.

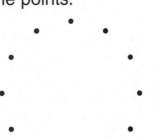

Which letter names the right angle? _____

6. Divide.

$30 \div 6 =$ _____

$12 \div 4 =$ _____

$20 \div 5 =$ _____

_____ $= 14 \div 7$

_____ $= 9 \div 3$

154   one hundred fifty-four

Lesson 6·13 Math Boxes

1. Solve.

2 × 2 = _____

5 × 5 = _____

3 × 3 = _____

4 × 4 = _____

2. Solve.

Double 3 _____

Double 30 _____

Double 300 _____

Double 7 _____

Double 70 _____

Double 700 _____

3. Solve.

5 × 4 = _____

2 × 7 = _____

_____ = 3 × 10

_____ = 7 × 10

3 × 5 = _____

4. Write 4 multiplication facts you need to practice.

_____

_____

_____

_____

5. Write 4 division facts you need to practice.

_____

_____

_____

_____

6. Complete the Fact Triangle. Write the fact family.

_____

_____

_____

_____

one hundred fifty-five

Date | Time

Notes

Date   Time

Notes

Date　　　　　　　　　　　　　　　Time

Notes

Date | Time

Notes

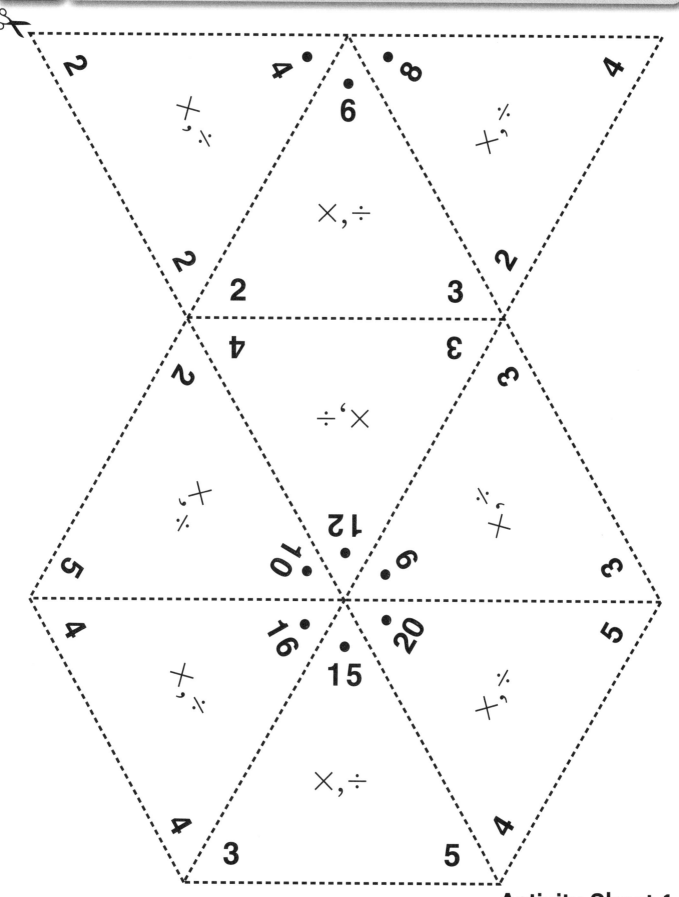

Activity Sheet 1

LESSON 4·6 Multiplication/Division Fact Triangles 2

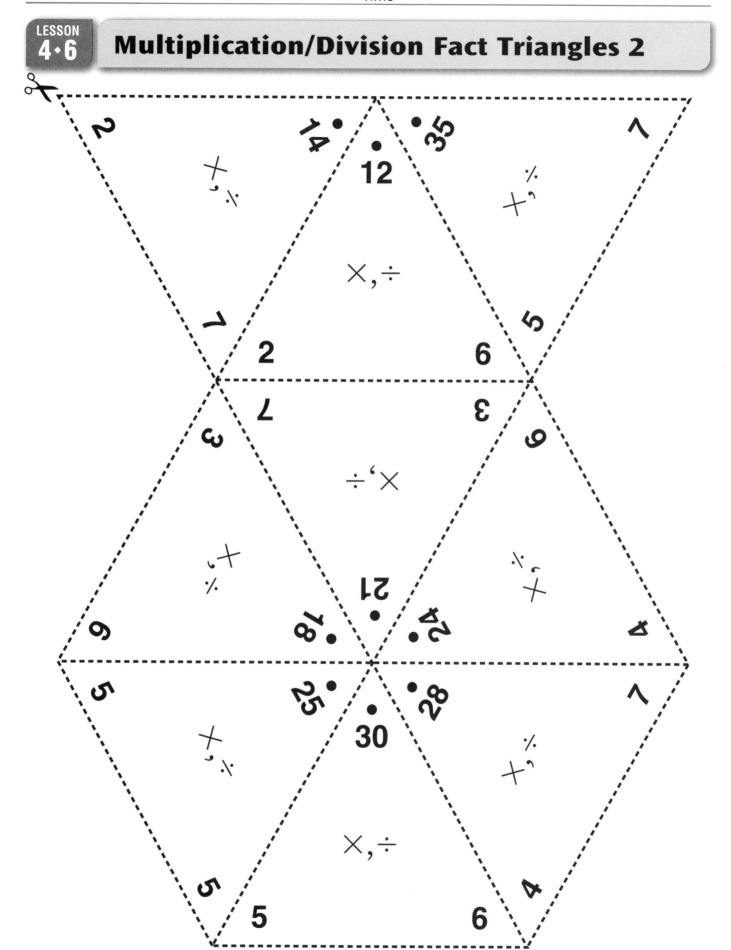

Activity Sheet 2

Date   Time

LESSON 7·2 ×, ÷ Fact Triangles 3

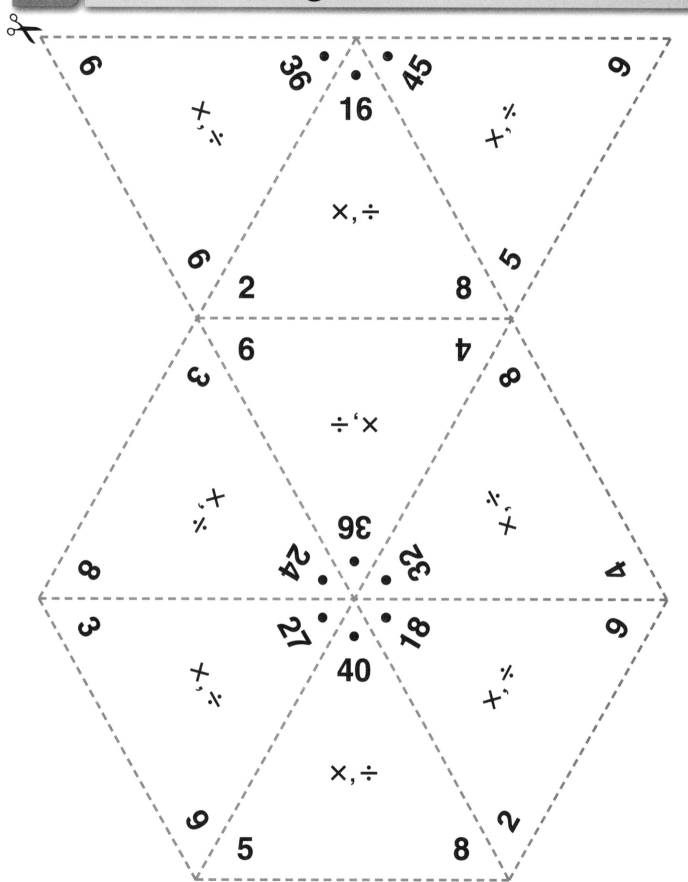

Activity Sheet 3

| Date | Time |

LESSON 7·2 ×, ÷ Fact Triangles 3

5 by 9

6 by 6

2 by 8

4 by 8

4 by 9

3 by 8

2 by 9

3 by 9

5 by 8

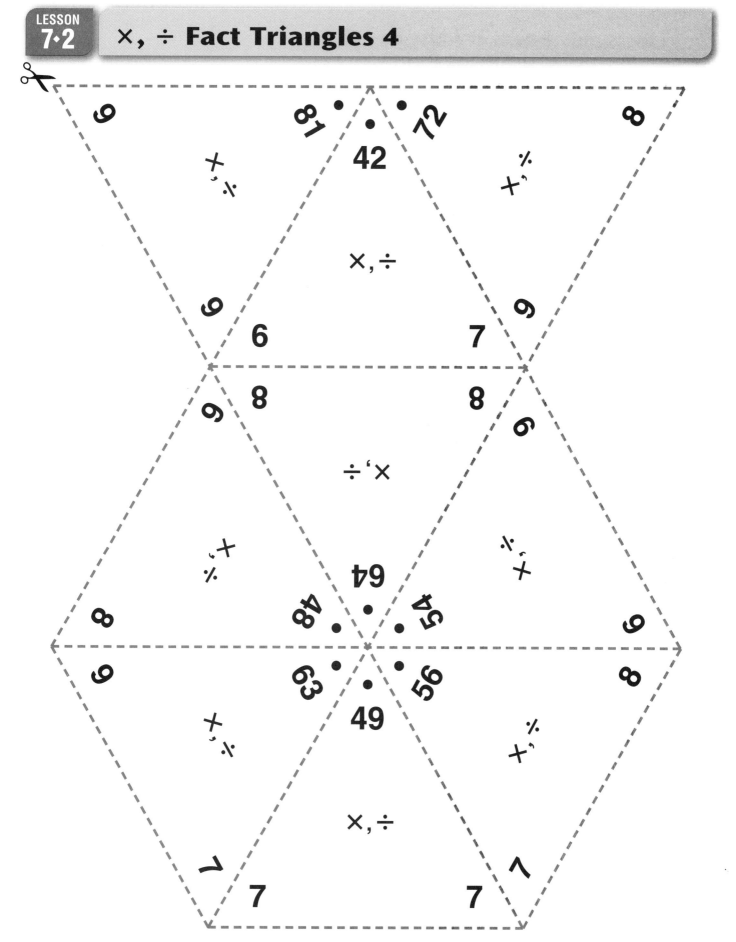

LESSON 7·2   ×, ÷ Fact Triangles 4

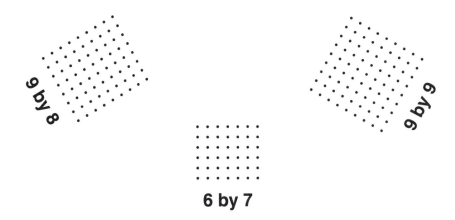

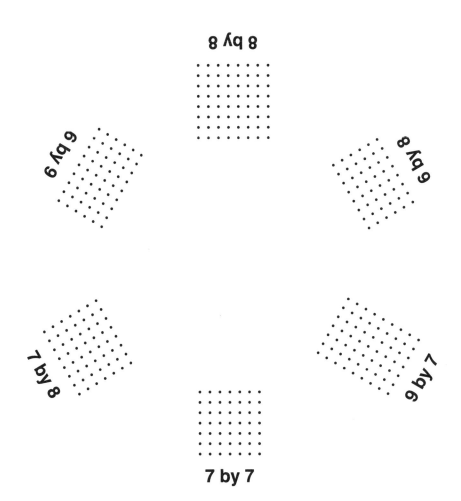

Everyday Mathematics®

The University of Chicago School Mathematics Project

Student Math Journal
Volume 2

Grade 3

McGraw Hill Wright Group

The University of Chicago School Mathematics Project (UCSMP)

Max Bell, Director, UCSMP Elementary Materials Component; Director, *Everyday Mathematics* First Edition
James McBride, Director, *Everyday Mathematics* Second Edition
Andy Isaacs, Director, *Everyday Mathematics* Third Edition
Amy Dillard, Associate Director, *Everyday Mathematics* Third Edition

Authors

Max Bell	Amy Dillard	Kathleen Pitvorec
Jean Bell	Robert Hartfield	Peter Saecker
John Bretzlauf	Andy Isaacs	
Mary Ellen Dairyko*	James McBride	

Third Edition only

Technical Art
Diana Barrie

Teachers in Residence
Lisa Bernstein, Carole Skalinder

Editorial Assistant
Jamie Montague Callister

Contributors
Carol Arkin, Robert Balfanz, Sharlean Brooks, James Flanders, David Garcia, Rita Gronbach, Deborah Arron Leslie, Curtis Lieneck, Diana Marino, Mary Moley, William D. Pattison, William Salvato, Jean Marie Sweigart, Leeann Wille

Photo Credits
©Willem Bosman/Shutterstock, p. vii *bottom;* ©Tim Flach/Getty Images, cover; Getty Images, cover, *bottom left;* ©iStockphoto, pp. iii, v; Royalty-free/Corbis, p. vii *top.*

www.WrightGroup.com

Copyright © 2007 by Wright Group/McGraw-Hill.

All rights reserved. Except as permitted under the United States Copyright Act, no part of this publication may be reproduced or distributed in any form or by any means, or stored in a database or retrieval system, without the prior written permission from the publisher, unless otherwise indicated.

Printed in the United States of America.

Send all inquiries to:
Wright Group/McGraw-Hill
P.O. Box 812960
Chicago, IL 60681

ISBN 0-07-604568-4

13 14 15 CPC 12 11 10 09

The McGraw·Hill Companies

Contents

UNIT 7 — Multiplication and Division

Product Patterns	157
Math Boxes 7•1	158
Multiplication/Division Facts Table	159
Math Boxes 7•2	160
Multiplication Bingo	161
Multiplication/Division Practice	162
Math Boxes 7•3	163
Number Models with Parentheses	164
Math Boxes 7•4	165
Scoring 10 Basketball Points	166
Names with Parentheses	167
Math Boxes 7•5	168
Extended Multiplication and Division Facts	169
Math Boxes 7•6	170
Stock-up Sale Record	171
Math Boxes 7•7	172
Tens Times Tens	173
Math Boxes 7•8	174
National High/Low Temperatures Project	175
Temperature Ranges Graph	176
Math Boxes 7•9	178
Math Boxes 7•10	179

Contents **iii**

UNIT 8 Fractions

Fraction Review	180
Math Boxes 8•1	182
Drawing Blocks	183
Math Boxes 8•2	184
Fractions with Pattern Blocks	185
Dressing for the Party	188
Math Boxes 8•3	190
Fraction Number-Line Poster	191
Frames-and-Arrows Problems	192
Math Boxes 8•4	193
Table of Equivalent Fractions	194
Math Boxes 8•5	195
Math Boxes 8•6	196
More Than ONE	197
Math Boxes 8•7	199
Fraction Number Stories	200
Math Boxes 8•8	202
Math Boxes 8•9	203

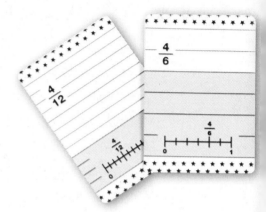

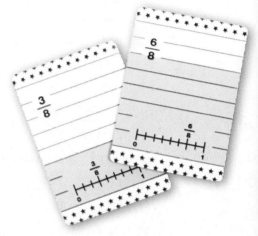

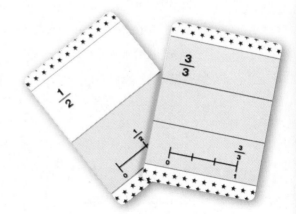

UNIT 9   Multiplication and Division

Adult Weights of North American Animals **204**
Multiples of 10, 100, and 1,000 .... **206**
Math Boxes 9♦1 .... **207**
Mental Multiplication **208**
Number Stories **209**
Math Boxes 9♦2 .... **210**
Array Multiplication 1 .... **211**
Geoboard Areas **212**
Math Boxes 9♦3 .... **213**
Using the Partial-Products Algorithm **214**
Measures **215**
Math Boxes 9♦4 .... **216**
Shopping at the Stock-Up Sale **217**
Math Boxes 9♦5 .... **218**
Factor Bingo Game Mat **219**
Using the Partial-Products Algorithm **220**
Math Boxes 9♦6 .... **221**
Sharing Money **222**
Math Boxes 9♦7 .... **223**
Division with Remainders **224**
Math Boxes 9♦8 .... **225**
Lattice Multiplication **226**
Lattice Multiplication Practice **227**
Math Boxes 9♦9 .... **228**
Array Multiplication 2 .... **229**
Array Multiplication 3 .... **230**
Sharing Money **231**
Math Boxes 9♦10 .... **232**
Multiplication with Multiples of 10 .... **233**
Math Boxes 9♦11 .... **234**
2-Digit Multiplication **235**
Math Boxes 9♦12 .... **236**
Number Stories with Positive
   and Negative Numbers **237**
Math Boxes 9♦13 .... **238**
Math Boxes 9♦14 .... **239**

Contents   **v**

UNIT 10 Measurement and Data

Review: Units of Measure	240
Weight and Volume	241
Multiplication Practice	242
Math Boxes 10•1	243
Volume of Boxes	244
Math Boxes 10•2	245
Various Scales	246
Reading Scales	247
Math Boxes 10•3	248
Math Boxes 10•4	249
Units of Measure	250
Body Measures	251
Math Boxes 10•5	252
A Mean, or Average, Number of Children	253
A Mean, or Average, Number of Eggs	254
Math Boxes 10•6	255
Finding the Median and the Mean	256
Math Boxes 10•7	257
Calculator Memory	258
Measurement Number Stories	259
Math Boxes 10•8	260
Frequency Table	261
Bar Graph	262
Math Boxes 10•9	263
Plotting Points on a Coordinate Grid	264
Math Boxes 10•10	265
Math Boxes 10•11	266

UNIT 11 Probability; Year-Long Projects, Revisited

Math Boxes 11•1 267
Math Boxes 11•2 268
Spinners 269
Estimate, Then Calculate 270
Math Boxes 11•3 271
Making Spinners 272
Math Boxes 11•4 274
Random-Draw Problems 275
Reading and Writing Numbers 276
Math Boxes 11•5 277
Math Boxes 11•6 278
Sunrise and Sunset Record 279
Length-of-Day Graph 280

Activity Sheets

Fraction Cards . . . . **Activity Sheet 5**
Fraction Cards . . . . **Activity Sheet 6**
Fraction Cards . . . . **Activity Sheet 7**
Fraction Cards . . . . **Activity Sheet 8**

Contents **vii**

Date _____ Time _____

LESSON 7·1 — Product Patterns

Part A

Math Message

Complete the facts.

1. $1 \times 1 =$ _____
2. $2 \times 2 =$ _____
3. $3 \times 3 =$ _____
4. $4 \times 4 =$ _____
5. $5 \times 5 =$ _____
6. $6 \times 6 =$ _____
7. $7 \times 7 =$ _____
8. $8 \times 8 =$ _____
9. $9 \times 9 =$ _____
10. $10 \times 10 =$ _____

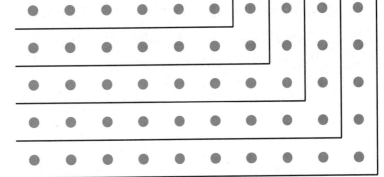

Part B

A Two's Product Pattern

Multiply. Look for patterns.

11. $2 \times 2 =$ _____

12. $2 \times 2 \times 2 =$ _____

13. $2 \times 2 \times 2 \times 2 =$ _____

14. $2 \times 2 \times 2 \times 2 \times 2 =$ _____

15. $2 \times 2 \times 2 \times 2 \times 2 \times 2 =$ _____

Try This

Use the Two's Product Pattern for Problems 11 through 15. Multiply.

16. $2 \times 2 \times 2 \times 2 \times 2 \times 2 \times 2 =$ _____

one hundred fifty-seven

Date _____ Time _____

LESSON 7·1 Math Boxes

1. This is a picture of a triangular pyramid. This shape has

    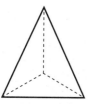

    ____ faces

    ____ edges

    ____ vertices

2. Draw an array with 25 Xs arranged in 5 rows.

    How many Xs in each row? ____

    Write a number model for the array.

    _____

3. Draw and label three parallel line segments. Draw and label a line that intersects all three line segments.

4. Fill in the circle next to the correct answer.

    777
    + 1,028
    ———

    Ⓐ 251   Ⓑ 1,751

    Ⓒ 1,795 Ⓓ 1,805

5. Complete the Fact Triangle. Write the fact family.

    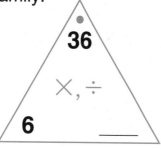
    36
    ×, ÷
    6

    ____ × ____ = ____

    ____ ÷ ____ = ____

6. Divide the rectangle into 4 equal parts.

158 one hundred fifty-eight

LESSON 7·2 Multiplication/Division Facts Table

×,÷	1	2	3	4	5	6	7	8	9	10
1	1	2	3	4	5	6	7	8	9	10
2	2	4	6	8	10	12	14	16	18	20
3	3	6	9	12	15	18	21	24	27	30
4	4	8	12	16	20	24	28	32	36	40
5	5	10	15	20	25	30	35	40	45	50
6	6	12	18	24	30	36	42	48	54	60
7	7	14	21	28	35	42	49	56	63	70
8	8	16	24	32	40	48	56	64	72	80
9	9	18	27	36	45	54	63	72	81	90
10	10	20	30	40	50	60	70	80	90	100

one hundred fifty-nine

Lesson 7·2 Math Boxes

1. Draw the lines of symmetry.

   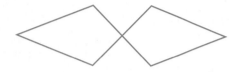

   There are ____ lines of symmetry.

2. Draw a trapezoid. Then draw a line segment that intersects the two parallel sides.

3. Solve.

   90 + 60 = _____

   _____ = 800 + 500

   1,700 − 900 = _____

   15,000 − 6,000 = _____

   _____ = 1,100 − 500

4. I have 3 vertices. All my sides are different lengths. I have one right angle. What am I?

   Fill in the circle for the best answer.

   ○ **A.** trapezoid

   ○ **B.** kite

   ○ **C.** hexagon

   ○ **D.** right triangle

5. Complete the number-grid puzzles.

   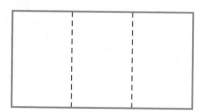

6. Shade $\frac{1}{3}$ of the rectangle.

160 one hundred sixty

Lesson 7·3 Multiplication Bingo

Read the rules for *Multiplication Bingo* on pages 293 and 294 in the *Student Reference Book*.

Write the list of numbers on each grid below.

List of numbers

1	9	18	30
4	12	20	36
6	15	24	50
8	16	25	100

Record the facts you miss.
Be sure to practice them!

_____

_____

_____

_____

_____

one hundred sixty-one **161**

Date **Time**

Lesson 7·3 Multiplication/Division Practice

Fill in the missing number in each Fact Triangle.
Then write the fact family for the triangle.

1.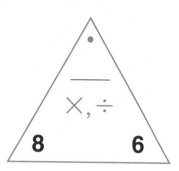

_____

_____

_____

2.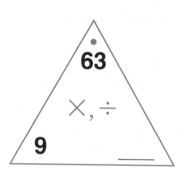

_____

_____

_____

Complete each puzzle.
Example:

×,÷	3	5
4	12	20
6	18	30

3.

×,÷	2	6
3		
6		

4.

×,÷	3	5
2		
8		

5.

×,÷	7	9
2		
5		

6.

×,÷		4
3	9	
4		

7.

×,÷		6
2		
	24	36

162 one hundred sixty-two

Date _____ Time _____

LESSON 7·3 Math Boxes

1. This is a picture of a cube. What do you know about this shape?

   _____
   _____
   _____

2. Draw an array of 27 Xs arranged in 3 rows.

   How many Xs in each row? _____

   Write a number model for the array.

   _____

3. Draw a ray $\overrightarrow{AB}$ that is parallel to the line $\overleftrightarrow{CD}$ and intersects the line segment $\overline{EF}$.

4. Write a number model for your ballpark estimate:

   _____

   Subtract and show your work:

   $\begin{array}{r} 926 \\ -538 \\ \hline \end{array}$

5. Fill in the blanks for this ×, ÷ puzzle.

×, ÷	5	
8		
	45	63

6. Divide each figure into 4 equal parts.

   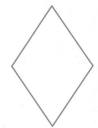

one hundred sixty-three **163**

Date _____ Time _____

LESSON 7·4 Number Models with Parentheses

Solve the number story. Then write a number model using parentheses.

1. Alexis scored 12 points, and Nehemie scored 6 points. If their team scored 41 points, how many points did the rest of the team score?

   _____

   Number model: _____

2. In a partner game, Quincy has 10 points, and Ellen has 14 points. They need 50 points to finish the game. How many more points are needed?

   _____

   Number model: _____

3. Quincy and Ellen earned 49 points but lost 14 points for a wrong move. They gained 10 points back. What was their score at the end of the round?

   _____

   Number model: _____

Complete these number sentences.

4. _____ = 18 − (9 + 5)   5. (75 − 29) + 5 = _____

6. _____ = 8 + (9 × 3)   7. 36 + (15 ÷ 3) = _____

Add parentheses to complete the number models.

8. 20 − 10 + 4 = 6   9. 20 − 10 + 4 = 14   10. 100 − 21 + 10 = 69

11. 100 − 21 + 10 = 89   12. 27 − 8 + 3 = 22   13. 18 = 6 + 3 × 4

164 one hundred sixty-four

Lesson 7·4

Math Boxes

1. Draw the lines of symmetry.

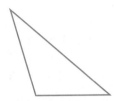

There are _____ lines of symmetry.

2. Draw a parallelogram. Label the vertices so $\overline{AB} \parallel \overline{CD}$. The symbol $\parallel$ means *is parallel to*.

3. Solve.

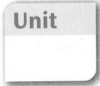

_____ = 400 + 800

3,000 + 7,000 = _____

90,000 − 20,000 = _____

4. Answer this riddle.

I have four sides. My opposite sides are equal in length. I have two pairs of parallel sides. I do not have any right angles.

What shape am I?

_____

5. Complete the number-grid puzzle.

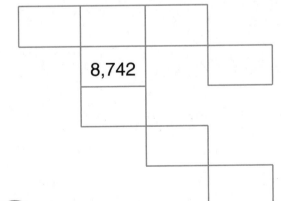

6. Divide the triangles into three equal groups.

one hundred sixty-five **165**

Date   **Time**

LESSON 7·5   Scoring 10 Basketball Points

Find different ways to score 10 points in a basketball game.

Number of 3-point Baskets	Number of 2-point Baskets	Number of 1-point Baskets	Number Models
2	2	0	$(2 \times 3) + (2 \times 2) + (0 \times 1) = 10$

LESSON 7·5 Names with Parentheses

Cross out the names that don't belong in each name-collection box.

1.
12

(3 × 3) + 3   3 × (3 + 3)

2 + (4 × 2)   (2 + 4) × 2

4 × (4 − 4)   (4 × 4) − 4

2.
20

2 × (9 + 1)   (2 × 9) + 1

30 − (5 × 2)   (30 − 5) × 2

(100 ÷ 10) + 10   100 ÷ (10 + 10)

Write names that contain parentheses in each name-collection box.

3.
16

4.
24

5. Write a parentheses problem. Describe how you solved the problem.

_____

_____

_____

_____

one hundred sixty-seven **167**

Date   **Time**

LESSON 7·5   Math Boxes

1. Solve.

   (6 × 3) + 2 = _____

   29 − (20 + 3) = _____

   _____ = 14 + (3 + 3)

   _____ = (5 × 5) − 6

2. Estimate. A bottle of milk costs $2.85. If Nan has $9, does she have enough money to buy 3 bottles of milk? (There is no tax.)

   _____

   Number model: _____

3. Fill in the oval next to the number model that best describes this array:

   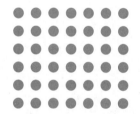

   ◯ 6 + 7 = 13   ◯ 6 × 7 = 42

   ◯ 7 × 7 = 49   ◯ 7 + 7 = 14

4. Fill in the fact triangle. Write the fact family.

   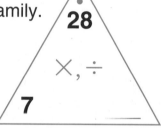

   ___ × ___ = ___

   ___ × ___ = ___

   ___ ÷ ___ = ___

   ___ ÷ ___ = ___

5. Solve the ×, ÷ puzzle. Fill in the blanks.

×, ÷	3	9
100		900
3,000		

6. Shade $\frac{1}{2}$ of the hexagon.

   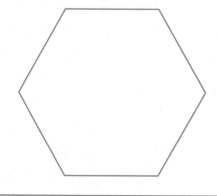

168   one hundred sixty-eight

Date _____ Time _____

LESSON 7·6 Extended Multiplication and Division Facts

Write the number of 3s in each number.

1. How many 3s in 30? _____
2. How many 3s in 300? _____
3. How many 3s in 3,000? _____
4. How many 3s in 12? _____
5. How many 3s in 120? _____
6. How many 3s in 1,200? _____

Solve each ×, ÷ puzzle. Fill in the blanks.

Example:

×, ÷	300	2,000
2	600	4,000
3	900	6,000

7.

×, ÷	60	300
4		
5	300	

Try This

8.

×, ÷	4	5
200		
8,000		

9.

×, ÷			1,000
3	1,500		
			6,000

10. Solve the number story.
    A 30-**minute** television program has
    ◆ two 60-**second** commercials at the beginning,
    ◆ two 60-**second** commercials at the end, and
    ◆ four 30-**second** commercials in the middle.

    a. How many **minutes** of commercials are there? _____
       (unit)

    b. How many **minutes** is the actual program? _____
       (unit)

    c. Number model: _____

one hundred sixty-nine **169**

Math Boxes

LESSON 7·6

1. Solve.

6 × 6 = _____

7 × 7 = _____

8 × 8 = _____

81 = _____ × _____

100 = _____ × _____

2. Fill in the missing whole number factors.

_____ × _____ = 14

28 = _____ × _____

32 = _____ × _____

_____ × _____ = 48

54 = _____ × _____

3. Add parentheses to complete the number models.

30 = 10 × 2 + 10

46 − 23 − 13 = 10

4 ÷ 2 + 6 = 8

4. Complete.

in
Rule
÷2
out

in	out
8	
16	
	10
50	

5. Solve.

6 × 10 = _____

6 × 30 = _____

50 × 6 = _____

_____ = 70 × 6

6 × 90 = _____

6. Color $\frac{1}{2}$ of the circle.

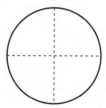

How many fourths are shaded?

_____ fourths

170 one hundred seventy

Date _____   Time _____

LESSON 7·7   Stock-up Sale Record

Use the items on pages 216 and 217 in your *Student Reference Book*.

Round 1:

Item to be purchased: _____

How many? _____

Regular or sale price? _____

Price per item: _____

Estimated cost: _____

Round 2:

Item to be purchased: _____

How many? _____

Regular or sale price? _____

Price per item: _____

Estimated cost: _____

Round 3:

Item to be purchased: _____

How many? _____

Regular or sale price? _____

Price per item: _____

Estimated cost: _____

Round 4:

Item to be purchased: _____

How many? _____

Regular or sale price? _____

Price per item: _____

Estimated cost: _____

Round 5:

Item to be purchased: _____

How many? _____

Regular or sale price? _____

Price per item: _____

Estimated cost: _____

Round 6:

Item to be purchased: _____

How many? _____

Regular or sale price? _____

Price per item: _____

Estimated cost: _____

Lesson 7·7 **Math Boxes**

1. Complete the number models.

   (49 − 19) − 8 = _____

   (56 − 14) × 2 = _____

   48 − (19 − 8) = _____

   56 − (14 − 2) = _____

2. Estimate: About how many dollars will Stephen need to buy 4 stopwatches for $12.89 each? (There is no tax.)

   Number model:

   _____

   He will need about

   $ _____

3. Solve. Write a number model.

groups	children per group	children in all
6	7	

   Number model

   _____

4. Complete the extended fact triangle. Write the fact family.

   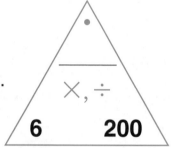

   ___ × ___ = ___

   ___ × ___ = ___

   ___ ÷ ___ = ___

   ___ ÷ ___ = ___

5. Solve the ×, ÷ puzzle. Fill in the blanks.

×, ÷		
		2,000
4	1,200	
		10,000

6. Shade $\frac{1}{2}$ of the balloons.

   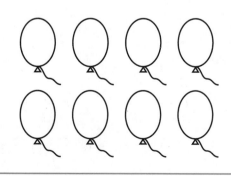

172 one hundred seventy-two

LESSON 7·8 Tens Times Tens

Math Message
Write the dollar values.

1. 10 $10 = $ _____
2. 100 $10 = $ _____
3. 1,000 $10 = $ _____
4. 10 $100 = $ _____
5. 100 $100 = $ _____
6. 1,000 $100 = $ _____

Solve each ×, ÷ puzzle. Fill in the blanks.

7.

×, ÷	10	100
1		
10		

8.

×, ÷	4	30
20		
6		

Try This

9.

×, ÷	40	60
20		
80		

10.

×, ÷		
3	150	
70		560

Multiply.

11. 5 × 90 = _____
12. _____ = 70 × 4
13. 7 × _____ = 420
14. _____ × 90 = 540
15. 10 × 70 = _____
16. 80 × 60 = _____
17. _____ = 30 × 50
18. _____ × 600 = 6,000

Lesson 7·8 Math Boxes

1. Solve.

    49 ÷ 7 = _____

    81 ÷ 9 = _____

    _____ = 64 ÷ 8

    6 = 36 ÷ _____

    _____ ÷ 5 = 5

2. Find the missing factors.

    36 = _____ × _____

    56 = _____ × _____

    _____ × _____ = 24

    _____ × _____ = 42

    18 = _____ × _____

3. Find the missing numbers, and add parentheses to make the number models true.

    3 × _____ + 5 = 29

    5 × 3 + _____ = 19

    25 = _____ × 7 − 3

4. Complete.

    in ↓
    Rule
    ÷12
    out

inches	feet
12	1
36	
	4
	2
60	

5. Solve.

    $\begin{array}{r}8\\ \times 5\\ \hline\end{array}$ $\begin{array}{r}80\\ \times 5\\ \hline\end{array}$ $\begin{array}{r}800\\ \times 5\\ \hline\end{array}$

    $\begin{array}{r}8{,}000\\ \times 5\\ \hline\end{array}$ $\begin{array}{r}5{,}000\\ \times 8\\ \hline\end{array}$

6. Fill in the oval in front of the fraction that does NOT represent the picture.

    ○ $\frac{6}{12}$

    ○ $\frac{3}{8}$

    ○ $\frac{1}{2}$

    ○ $\frac{2}{4}$

Date　　　　　　　　　　Time

LESSON 7·8　National High/Low Temperatures Project

Date	Highest Temperature (maximum)		Lowest Temperature (minimum)		Difference (range)
	Place	Temperature	Place	Temperature	
		°F		°F	°F
		°F		°F	°F
		°F		°F	°F
		°F		°F	°F
		°F		°F	°F
		°F		°F	°F
		°F		°F	°F
		°F		°F	°F
		°F		°F	°F
		°F		°F	°F
		°F		°F	°F
		°F		°F	°F
		°F		°F	°F
		°F		°F	°F
		°F		°F	°F
		°F		°F	°F
		°F		°F	°F
		°F		°F	°F
		°F		°F	°F
		°F		°F	°F
		°F		°F	°F

one hundred seventy-five **175**

Date _____ Time _____

LESSON 7·8 Temperature Ranges Graph

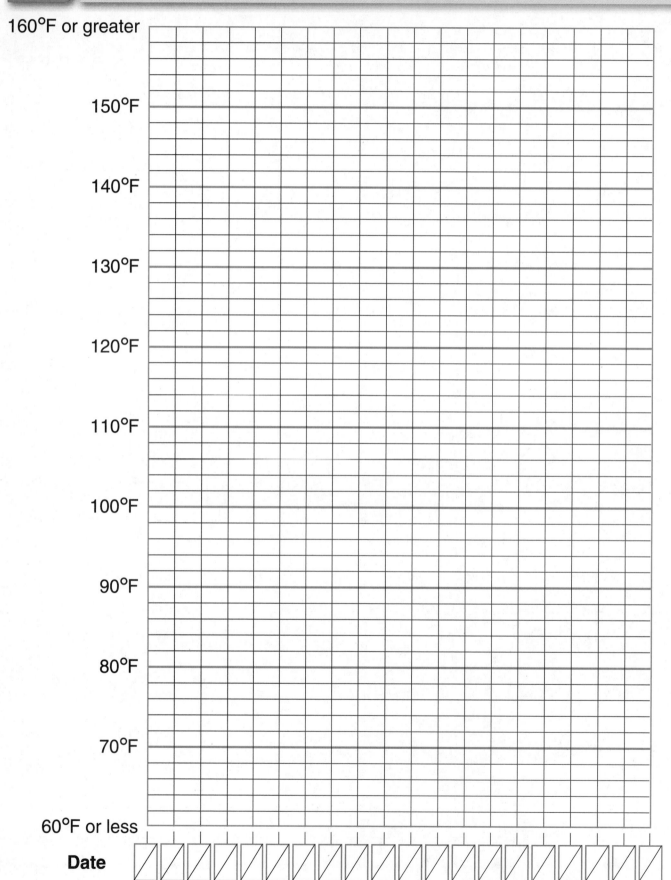

176 one hundred seventy-six

LESSON 7·8

Temperature Ranges Graph *continued*

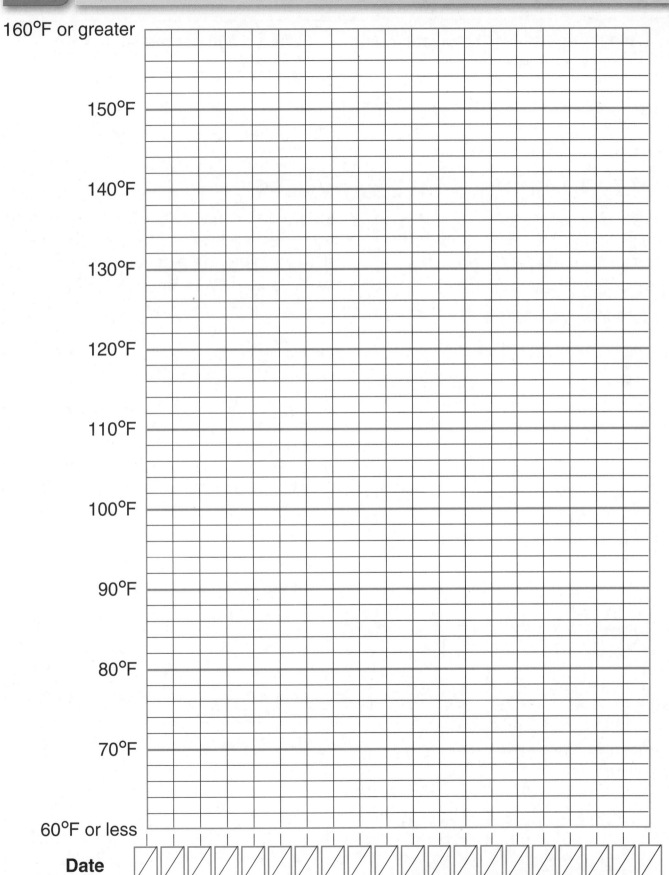

Math Boxes

LESSON 7·9

1. Add parentheses to complete the number models.

   $14 - 7 \times 2 = 14$

   $3 \times 6 + 2 = 24$

   $7 = 6 + 15 \div 3$

   $9 \times 5 + 3 = 72$

2. Write a number model for your ballpark estimate:

   _____

   Subtract. Show your work.

   900
   -799
   $\overline{}$

3. Solve. Show your work.

   7 cartons
   6 donuts per carton

   How many donuts in all?

   _____ donuts

4. Complete the extended Fact Triangle. Write the fact family.

   _____

   _____

   _____

   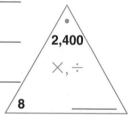

5. Three of the names do not belong in this name-collection box. Cross them out.

   4,000

   $8 \times 5,000$ $(500 \times 5) - 500$

   $5,000 - (5 \times 200)$

   $(200 \times 4) \times 5$ $2 \times 2,000$

   $(200 \div 4) \times 8$ $1,000 \times 4$

   $8,000 \div 2$ $(2 \times 2) \times 1,000$

   $(200 + 200) \times 10$

6. Divide the triangles into 2 equal groups.

   △ △ △ △

   △ △ △ △

   △ △ △ △

Date _____ Time _____

LESSON 7·10 Math Boxes

1. Circle the pictures in which $\frac{1}{2}$ is shaded.

2. True or false?

$\frac{1}{3}$ of the squares are shaded.

_____

3. Circle $\frac{7}{18}$.

What fraction of dots is not circled?

_____

4. Color $\frac{3}{8}$ yellow and $\frac{4}{8}$ red.

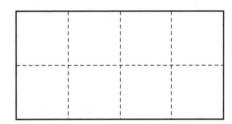

How much is not colored?

_____

5. Write a number or a fraction.

There are ____ quarters in one dollar.

One quarter = ____ of a dollar.

There are ____ dimes in one dollar.

One dime = ____ of a dollar.

6. Divide the rhombus into 4 equal parts.

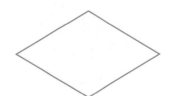

What fraction of the rhombus is each part?

_____

one hundred seventy-nine **179**

Date _____ Time _____

LESSON 8·1 Fraction Review

Math Message

1. Draw an X through $\frac{2}{3}$ of the circles. ○ ○ ○ ○ ○ ○

Label each picture with one of the following numbers: 0, $\frac{0}{4}$, $\frac{1}{4}$, $\frac{1}{2}$, $\frac{2}{4}$, or $\frac{3}{4}$.

  2.   3.   4.   5.

$\frac{4}{4}$ or 1   _____   _____   _____   _____

Each whole figure represents ONE.
Write a fraction that names each region inside the figure.

6. 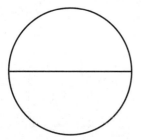 7. 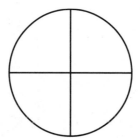 8.

9. 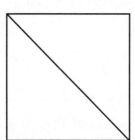 10.  11.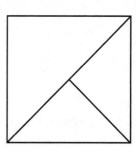

Try This

Each whole figure represents ONE. Write a fraction that names each region inside the figure.

12. 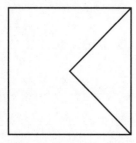 13.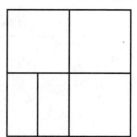

180 one hundred eighty

Date _____ Time _____

LESSON 8·1 Fraction Review continued

You need at least 25 pennies or other counters to help you solve these problems. Share solution strategies with others in your group.

Unit

counters

14. a. Show $\frac{1}{4}$ of a set of 8 counters. How many counters is that? _____
    b. Show $\frac{2}{4}$ of the set. How many counters? _____
    c. Show $\frac{3}{4}$ of the set. How many counters? _____

15. a. Show $\frac{1}{3}$ of a set of 12 counters. How many counters is that? _____
    b. Show $\frac{2}{3}$ of the set. How many counters? _____
    c. Show $\frac{3}{3}$ of the set. How many counters? _____

16. a. Show $\frac{1}{5}$ of a set of 15 counters. How many counters is that? _____
    b. Show $\frac{4}{5}$ of the set. How many counters? _____

17. Show $\frac{3}{4}$ of a set of 20 counters. How many counters? _____

18. Show $\frac{2}{3}$ of a set of 18 counters. How many counters? _____

19. Five counters is $\frac{1}{5}$ of a set. How many counters are in the whole set? _____

20. Six counters is $\frac{1}{3}$ of a set. How many counters are in the whole set? _____

Try This

21. Twelve counters is $\frac{3}{4}$ of a set. How many counters are in the complete set? _____

22. Pretend that you have a set of 15 cheese cubes. What is $\frac{1}{2}$ of that set? Use a fraction or decimal in your answer. _____

one hundred eighty-one **181**

Date _____ Time _____

LESSON 8·1 Math Boxes

1. Double each amount.

$0.25 _____

$0.50 _____

$0.75 _____

$1.25 _____

$5.00 _____

2. Fill in the missing numbers.

×, ÷	700	60
8		
	4,900	

3. This drawing shows a rectangular prism.

It has _____ faces.

It has _____ edges.

It has _____ vertices.

SRB 115

4. Complete the number models.

(4 + 3) − 2 = _____

10 = 6 + (2 + _____)

_____ = 3 × (9 − 0)

(5 × 5) − 4 = _____

SRB 16

5. Shade $\frac{3}{8}$ of the circle.

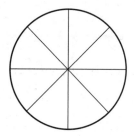

What fraction is *un*shaded? _____

SRB 22 23

6. 9 cups. 9 ice cubes per cup. How many ice cubes in all? Fill in the circle for the best answer.

◯ **A** 18 ice cubes

◯ **B** 81 ice cubes

◯ **C** 90 ice cubes

◯ **D** 99 ice cubes

SRB 66 67

LESSON	**Drawing Blocks**
8·2	

Color the blocks in the bags blue. Then fill in the blanks by answering this question: How many red blocks would you put into each bag?

1. If I wanted to be sure of taking out a blue block, I would put in _____ red block(s).

2. If I wanted to have an equal chance of taking out red or blue, I would put in _____ red block(s).

3. If I wanted to be more likely to take out blue than red, I would put in _____ red block(s).

4. If I wanted to take out a red block about 3 times as often as a blue one, I would put in _____ red block(s).

5. If I wanted to take out a red block about half as often as a blue one, I would put in _____ red block(s).

6. If I wanted to take out a red block about $\frac{1}{3}$ of the time, I would put in _____ red block(s).

Try This

7. If I wanted to take out a red block about $\frac{2}{3}$ of the time, I would put in _____ red block(s).

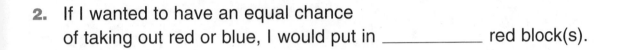
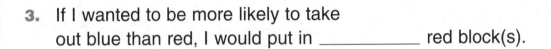
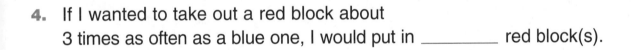
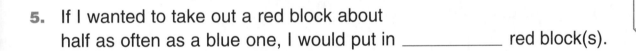
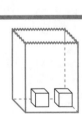

one hundred eighty-three

Date _____ Time _____

LESSON 8·2 Math Boxes

1. Shade $\frac{3}{7}$ of the books.

2. Complete the bar graph.

   Max swam 5 laps.

   Colin swam 3 laps.

   Miles swam 6 laps.

   Median number of laps: _____

   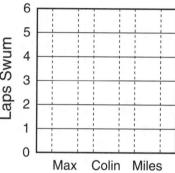

3. Circle $\frac{5}{10}$ of the collection of triangles.

   Write 2 names for the fraction that is left.

   _____ and _____

4. What is the missing factor? Fill in the circle for the best answer.

   $6 \times$ _____ $= 3{,}600$

   Ⓐ 6

   Ⓑ 60

   Ⓒ 600

   Ⓓ 6,000

5. Flip 1 penny and 1 nickel. Show all possible outcomes. Use Ⓟ and Ⓝ to show the coins. Use H for HEADS and T for TAILS.

6. How much do six 300-pound dolphins weigh?

dolphins	pounds per dolphin	pounds in all

   Answer: _____

   Number model:

   _____

184   one hundred eighty-four

Date _____ Time _____

LESSON 8·3 Fractions with Pattern Blocks

Exploration A

Work with a partner.

Materials ☐ pattern blocks
☐ Pattern-Block Template

Part 1

Cover each shape with green △ pattern blocks. What fractional part of each shape is 1 green pattern block? Write the fraction under each shape.

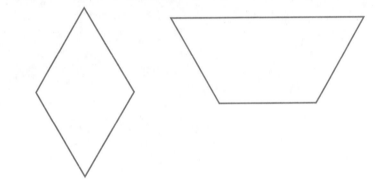

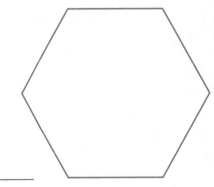

_____   _____   _____

Part 2

Cover each shape with green △ pattern blocks. What fractional part of each shape are 2 green pattern blocks? Write the fraction next to each shape.

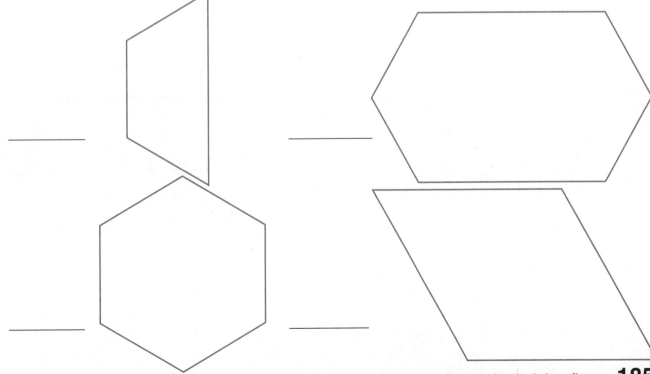

_____   _____

_____   _____

one hundred eighty-five **185**

LESSON 8·3 Fractions with Pattern Blocks continued

Part 3

Cover each shape with blue pattern blocks. What fractional part of each shape is 1 blue pattern block? Write the fraction under each shape. If you can't cover the whole shape, cover as much as you can. *Think:* Is there another block that would cover the rest of the shape?

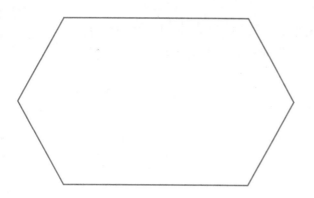

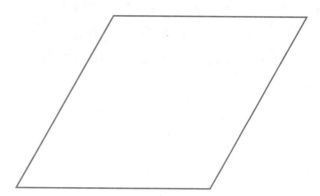

———   ———

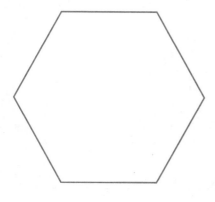

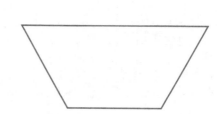

———   ———

| Date | Time |

LESSON 8·3 Fractions with Pattern Blocks continued

Try This

Part 4

Cover each shape with blue ◊ pattern blocks. What fractional part of each shape would 2 blue pattern blocks cover? Write the fraction under each shape.

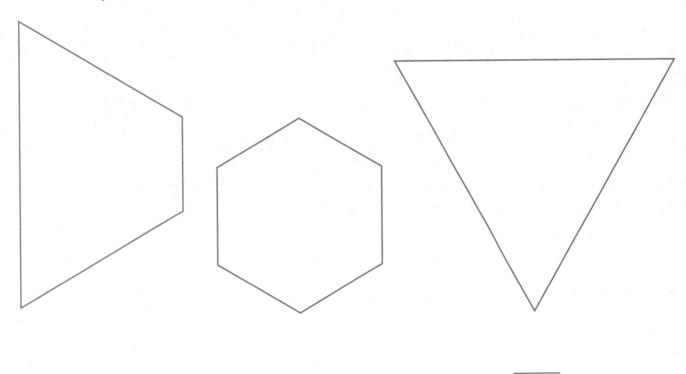

Part 5

Use your Pattern-Block Template to show how you divided the shapes in each section. *Remember:* The number *under* the fraction bar names the number of equal parts into which the whole shape is divided.

Follow-Up

Get together with the rest of the group.

◆ Compare your answers.

◆ Use the blocks to check your answers.

◆ Decide whether more than one fraction can be correct.

one hundred eighty-seven **187**

Date _____ Time _____

LESSON 8·3 Dressing for the Party

Exploration C

Work in a group of four.

Materials
- ☐ *Math Masters,* p. 244 (Pants and Socks Cutouts)
- ☐ scissors
- ☐ tape
- ☐ blue, red, green, and black crayons or coloring pencils

Problem

Pretend that you have 4 pairs of pants: blue, red, green, and black. You also have 4 pairs of socks: blue, red, green, and black. You have been invited to a party. You need to choose a pair of pants and a pair of socks to wear. (Of course, both socks must be the same color.) For example, the pants could be blue and both socks black.

How many different combinations of pants and socks are possible?

Strategy

Use the cutouts on *Math Masters,* page 244, and crayons to help you answer the question.

Before you answer the question, decide on a way for your group to share the following work.

◆ Color the pants in the first row blue.

◆ Color the pants in the second row red.

◆ Color the pants in the third row green and those in the fourth row black.

◆ Color the socks in the same way.

◆ Cut out each pair of pants and each pair of socks.

◆ Tape together pairs of pants and pairs of socks to show different outfits. Check that you have only one of each outfit.

188 one hundred eighty-eight

LESSON 8·3   **Dressing for the Party** *continued*

1. How many different combinations of pants and socks did your group find? _____

2. Is this all of the possible combinations? _____

3. How do you know?

   _____
   _____
   _____
   _____

4. How did your group divide up the work?

   _____
   _____
   _____
   _____

5. How did your group solve the problem?

   _____
   _____
   _____
   _____

Date Time

LESSON 8·3 Math Boxes

1. Double the amounts.

$1.10 _____

$2.50 _____

$10.50 _____

$12.50 _____

$25.00 _____

2. Solve.

5 × 9 = _____

5 × 90 = _____

5 × 900 = _____

_____ = 3 × 8

_____ = 30 × 80

_____ = 300 × 80

3. This drawing shows a square pyramid.

It has ____ faces.

It has ____ edges.

It has ____ vertices.

What is the shape of its base?

_____

4. Put in the parentheses needed to complete the number models.

31 = 3 + 7 × 4

40 = 3 + 7 × 4

4 × 8 + 2 × 2 = 36

4 × 8 + 2 × 2 = 80

5. Color $\frac{2}{5}$ of the rectangle.

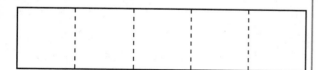

What fraction is *not* colored? _____

6. 12 apples per bag.

How many apples in 3 bags?

_____

How many apples in 4 bags?

_____

190   one hundred ninety

Date **Time**

LESSON 8·4 Fraction Number-Line Poster

1 Whole
Halves
Fourths
Eighths
Thirds
Sixths

one hundred ninety-one **191**

| Date | | Time |

Lesson 8·4 Frames-and-Arrows Problems

Solve each Frames-and-Arrows problem. Use your Fraction Number-Line Poster on *Math Journal 2,* page 191 for Problems 1 and 2.

1.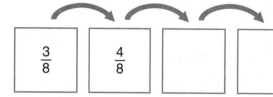

   Rule: $\frac{1}{8}$ more

   $\frac{3}{8}$ | $\frac{4}{8}$ | ___ | ___ | $\frac{7}{8}$ | $\frac{8}{8}$

2.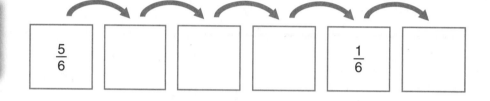

   Rule: $-\frac{1}{6}$

   $\frac{5}{6}$ | ___ | ___ | ___ | $\frac{1}{6}$ | ___

3.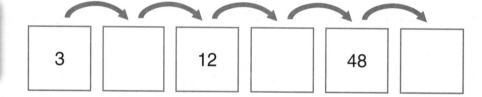

   Rule: ___

   3 | ___ | 12 | ___ | 48 | ___

4.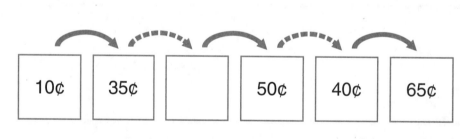

   10¢ | 35¢ | ___ | 50¢ | 40¢ | 65¢

Try This

5.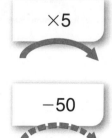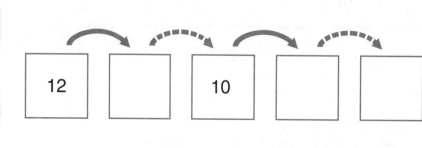

   Rules: ×5, −50

   12 | ___ | 10 | ___ | ___

Date _____ **Time** _____

Lesson 8·4 Math Boxes

1. Shade $\frac{7}{10}$ of the hats.

2. Use the bar graph.

Maximum: _____ number of laps

Minimum: _____ number of laps

Range: _____ number of laps

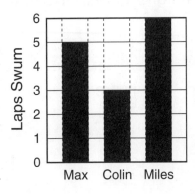

3. Suppose you like pizza and are very hungry. Would you rather have $\frac{4}{5}$ of a pizza or $\frac{8}{10}$ of a pizza?

_____

Why? _____

_____

_____

4. Fill in the missing numbers.

×, ÷		600
50	1,500	
		42,000

5. True or false? There is an equal chance of taking a B or an R block out of the bag.

_____

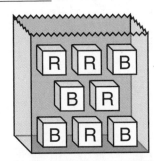

6. How much do seven packs of pencils cost if each pack costs $0.80?

packs of pencils	cost per pack	cost in all

Answer: _____

Number model: _____

one hundred ninety-three **193**

LESSON 8·5 Table of Equivalent Fractions

Use your deck of Fraction Cards to find equivalent fractions. Record them in the table.

Fraction	Equivalent Fractions
$\frac{0}{2}$	
$\frac{1}{2}$	
$\frac{2}{2}$	
$\frac{1}{3}$	
$\frac{2}{3}$	
$\frac{1}{4}$	
$\frac{3}{4}$	
$\frac{1}{5}$	
$\frac{4}{5}$	
$\frac{1}{6}$	
$\frac{5}{6}$	

Describe any patterns you see.

_____

_____

_____

Date _____ Time _____

LESSON 8·5 Math Boxes

1. In the number 3.514:

   the 3 means __3 ones__

   the 1 means _____

   the 5 means _____

   the 4 means _____

2. Which is true? Fill in the circle for the best answer.

   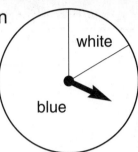

   The spinner is

   Ⓐ unlikely to land on blue

   Ⓑ less likely to land on white

   Ⓒ equally likely to land on blue or white

   Ⓓ likely to land on green

3. Write 4 fractions equivalent to $\frac{1}{2}$.

   _____ _____

   _____ _____

4. Use a straightedge. Draw the other half of the symmetric shape.

   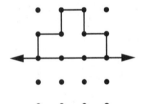

5. 9 children share 18 candies. How many candies per child?

   _____

   How many candies left over?

   _____

   16 books in all. 3 books per shelf.

   How many shelves? _____
   How many books left over?

   _____

6. Solve.

   54 ÷ 9 = _____

   27 ÷ 3 = _____

   _____ = 36 ÷ 6

   _____ = 64 ÷ 8

   45 ÷ 5 = _____

LESSON 8·6 Math Boxes

1. Write 4 fractions equivalent to $\frac{1}{4}$.

   _____   _____

   _____   _____

2. Complete.

   _____ hours = 1 day

   12 hours = _____ day

   _____ weeks = 21 days

   _____ minutes = $\frac{1}{2}$ hour

   15 minutes = _____ hour

3. If I wanted to have an equal chance of taking out a circle or a square, I would put in

   _____ circle(s).

   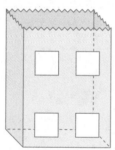

4. Draw a set of 12 Xs. Circle 9 of them. What fraction of the whole set is the 9 Xs?

   _____

5. Solve. Fill in the circle that shows the best answer.

   (2 × 90) + 7 = _____

   Ⓐ 98

   Ⓑ 99

   Ⓒ 187

   Ⓓ 194

6. Solve. Use your calculator. Pretend the division key is broken.

   Christopher and Rochelle are packing 212 cookies in boxes. Each box holds 20 cookies. How many **full** boxes can they pack?

   Answer: _____
   (unit)

Date	Time

LESSON 8·7 More Than ONE

Use the circles that you cut out for the Math Message.

1. Glue 3 halves into the two whole circles.

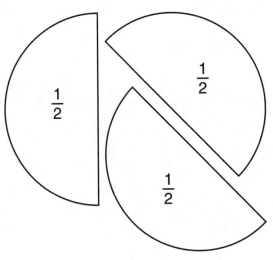

3 halves or $\frac{3}{2}$

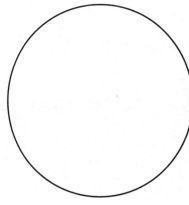

$1\frac{1}{2}$ or one and 1 half

2. Glue 6 fourths into the two whole circles. Fill in the missing digits in the question, the fraction, and the mixed number.

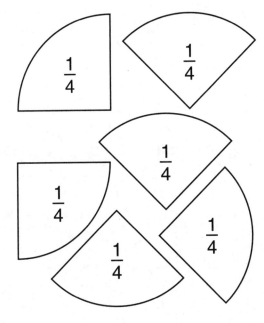

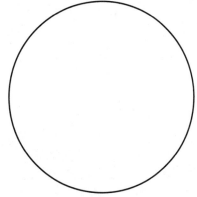

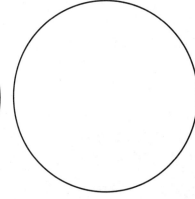

How many fourths? _____ fourths

Write the fraction:

Write the mixed number: 1

one hundred ninety-seven 197

Date	Time

LESSON 8·7 More Than ONE continued

3.

How many fourths? _____ fourths

Write the fraction: ☐/☐

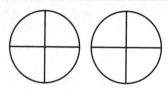

Color 5 fourths.

Write the mixed number: 1 ☐/☐

4.

How many thirds? _____ thirds

Write the fraction: ☐/☐

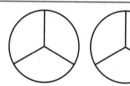

Color 5 thirds.

Write the mixed number: 1 ☐/☐

5.

How many fifths? _____ fifths

Write the fraction: ☐/☐

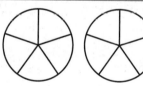

Color 8 fifths.

Write the mixed number: ☐ ☐/☐

6.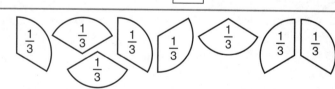

How many thirds? _____ thirds

Write the fraction: ☐/☐

Color 8 thirds.

Write the mixed number: ☐ ☐/☐

Date _____ Time _____

Lesson 8·7 Math Boxes

1. In the number 56.714:

the 7 means __7 tenths__

the 6 means _____

the 4 means _____

the 5 means _____

the 1 means _____

SRB 35

2. On which color is the spinner most likely to land? _____

Least likely to land? _____

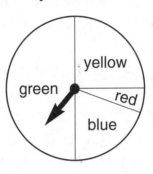

SRB 92 93

3. Circle the fractions that are equivalent to $\frac{1}{3}$.

$\frac{1}{8}$ $\frac{2}{6}$ $\frac{4}{12}$

$\frac{6}{9}$ $\frac{5}{15}$ $\frac{3}{9}$

SRB 30

4. Use a straightedge. Draw the other half of the symmetric shape.

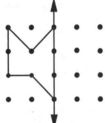

SRB 122 123

5. Share $3.75 equally among 3 people.

Each person gets $_____.

Share $10.00 equally among 4 people.

Each person gets $_____.

6. Solve.

6 × 8 = _____

9 × 9 = _____

7 × 7 = _____

_____ = 8 × 9

_____ = 4 × 8

SRB 52 53

one hundred ninety-nine **199**

LESSON 8·8 Fraction Number Stories

Solve these number stories. Use pennies, counters, or draw pictures to help you.

1. There are 8 apples in the package. Glenn did not eat any. What fraction of the package did Glenn eat?

   _____

2. Anik bought a dozen eggs at the supermarket. When he got home, he found that $\frac{1}{6}$ of the eggs were cracked. How many eggs were cracked?

   _____ eggs

3. Chante used $\frac{2}{3}$ of a package of ribbon to wrap presents. Did she use more or less than $\frac{3}{4}$ of the package?

   _____

4. I had 2 whole cookies. I gave you $\frac{1}{4}$ of 1 cookie. How many cookies did I have left?

   _____ cookies

5. There are 10 quarters. You have 3. I have 2. What fraction of the quarters do you have?

   _____

   What fraction of the quarters do I have?

   _____

   What fraction of the quarters do we have together?

   _____

6. One day, Edwin read $\frac{1}{3}$ of a book. The next day, he read another $\frac{1}{3}$ of the book. What fraction of the book had he read after 2 days?

   _____

   What fraction of the book did he have left to read?

   _____

7. Dorothy walks $1\frac{1}{2}$ miles to school. Jaime walks $1\frac{2}{4}$ miles to school. Who walks the longer distance?

   _____

8. Twelve children shared 2 medium-size pizzas equally. What fraction of 1 whole pizza did each child eat?

   _____

LESSON 8·8 Fraction Number Stories continued

9. Write a fraction story. Ask your partner to solve it.

_____

_____

_____

_____

_____

Draw eggs in each carton to show the fraction.

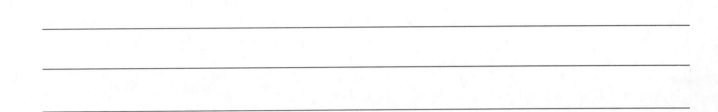

10. $\frac{6}{12}$

11. $\frac{4}{12}$

12. $\frac{3}{12}$

13. $\frac{1}{2}$

14. $\frac{3}{4}$

15. $\frac{1}{3}$

16. Julie drank $\frac{1}{4}$ of a glass of juice.

Draw an empty glass.

Shade in the glass to show how much juice is left. Write the fraction.

___ of the glass of juice is left.

two hundred one **201**

Date _____ Time _____

LESSON 8·8 Math Boxes

1. Draw two ways to show $\frac{2}{3}$.

2. 6 feet = _____ yards

   _____ feet = 18 inches

   $1\frac{1}{3}$ yards = _____ feet

   $1\frac{1}{2}$ yards = _____ inches

3. Use simple drawings to show all of the possible ways you can take 2 blocks from the bag.

   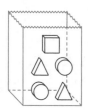

4. Tara frosted $\frac{4}{5}$ of the cupcakes. What fraction of the cupcakes is *not* frosted? _____

   Did she frost more or less than $\frac{1}{2}$ of the cupcakes? _____

   If there were 20 cupcakes in all, how many did she frost?

   _____

5. Show two ways a team can score 37 points in a football game.

7 points	6 points	3 points	2 points

   Write a number model:

   _____

   _____

6. Use your calculator. Pretend the division key is broken. Solve this problem.

   Will, Wes, Sam, and Ameer want to share $25 equally. How much money will each person get?

   Answer: _____

202 two hundred two

| Date | Time |

LESSON 8·9 Math Boxes

1. Solve.

   $7 \times 6 =$ _____

   $7 \times 60 =$ _____

   $7 \times 600 =$ _____

   _____ $= 8 \times 6$

   _____ $= 8 \times 60$

   _____ $= 8 \times 600$

2. Share $2.70 equally among 3 people.

   Each person gets $_____.

   Share $9 equally among 4 people.

   Each person gets $_____.

3. 30 is 10 times as much as _____.

   500 is _____ times as much as 5.

   _____ is 100 times as much as 80.

   40,000 is 1,000 times as much as _____.

4. 6 coats. 4 buttons per coat. How many buttons in all?

   _____ buttons

   Write a number model.

   _____

5. Draw a 4-by-7 array of Xs.

   How many Xs in all? _____

   Write a number model.

   _____

6. 18 books. 6 books per shelf.

   How many shelves? _____

   How many books left over? _____

   4 children share 13 marbles. How many marbles per child?

   _____

   How many marbles left over?

   _____

two hundred three **203**

LESSON 9·1 Adult Weights of North American Animals

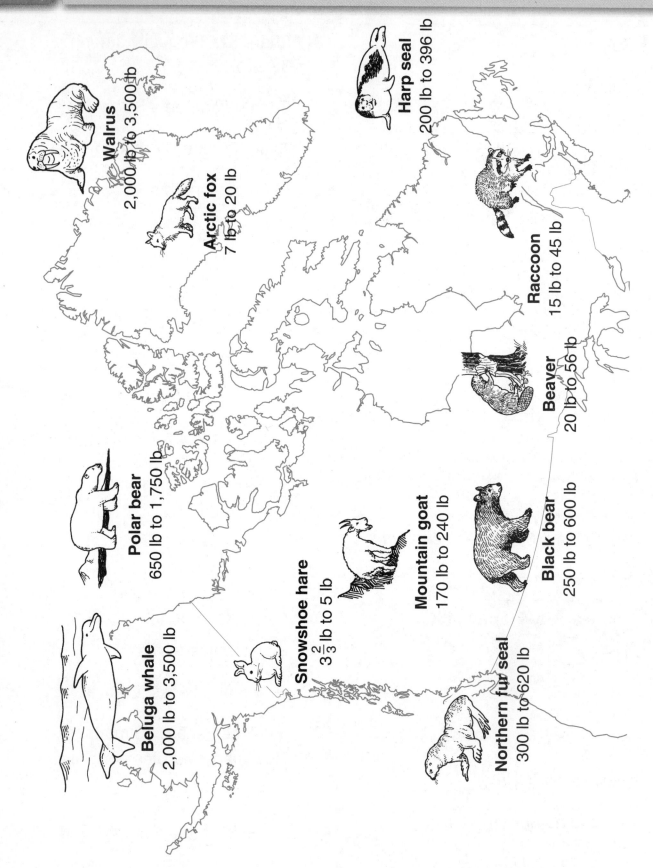

LESSON 9·1 **Adult Weights of North American Animals** *cont.*

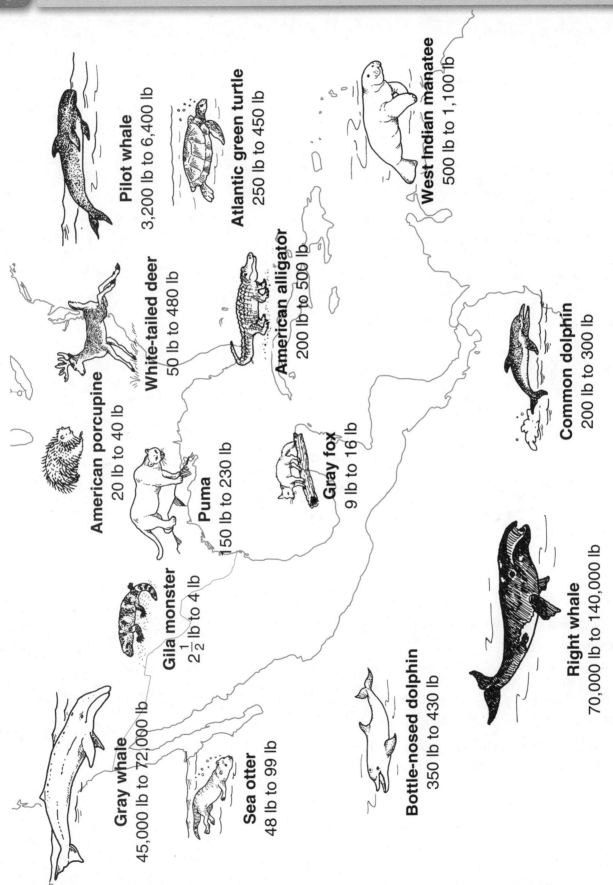

LESSON 9·1 Multiples of 10, 100, and 1,000

Solve each problem.

1. a. 7 [40s] = _____ b. 7 × 40 = _____

2. a. 6 [70s] = _____ b. 6 × 70 = _____

3. a. 60 [20s] = _____ b. 60 × 20 = _____

4. How many 50s are in 4,000? _____

5. How many 800s are in 2,400? _____

6. a. How many 3s are in 270? _____ b. _____ × 3 = 270

   c. 270 ÷ 3 = _____

7. a. 40 × 300 = _____ b. 12,000 ÷ 40 = _____

For Problems 8 through 10, use the information on pages 204 and 205.

8. a. Which animal might weigh about 20 times as much as a 30-pound raccoon? _____

   b. Can you name two other animals that might weigh 20 times as much as a 30-pound raccoon?
   _____

9. About how many 200-pound American alligators weigh about as much as a 3,200-pound beluga whale? _____

Try This

10. Which animal might weigh about 100 times as much as the combined weights of a 15-pound arctic fox and a 10-pound arctic fox? _____

206 two hundred six

Date _____ Time _____

LESSON 9·1 Math Boxes

1. If I wanted to take out a square about 4 times as often as a circle, I would put in _____ square(s).

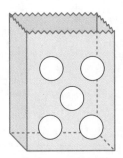

2. Put these numbers in order from smallest to largest.

    998,752   _____

    1,000,008 _____

    750,999   _____

    1,709,832 _____

3. Write equivalent fractions.

    $\frac{1}{2} = \frac{\Box}{\Box}$

    $\frac{1}{4} = \frac{\Box}{\Box}$

4. Pencils cost $1.99 for a package of 24. Estimate. About how much do 4 packages cost?

    _____

5. Use bills and coins.

    Share $45.90 equally among 3 people.

    Each gets $_____.

    Share $49.20 equally among 4 people.

    Each gets $_____.

6. Measure the line segment to the nearest $\frac{1}{2}$ inch.

    _____ about _____ in.

    Draw a line segment that is $1\frac{1}{2}$ inches long.

two hundred seven **207**

LESSON 9·2 **Mental Multiplication**

Solve these problems in your head. Use a slate and chalk, or pencil and paper, to help you keep track of your thinking. For some of the problems, you will need to use the information on journal pages 204 and 205.

1. Could 5 arctic foxes weigh 100 pounds? _____

   Less than 100 pounds? _____

   Explain the strategy you used.

   _____

   _____

2. Could 12 harp seals weigh more than 1 ton? _____ Less than 1 ton? _____

   Explain the strategy that you used.

   _____

   _____

3. How much do eight 53-pound white-tailed deer weigh? _____

   Explain the strategy that you used.

   _____

   _____

4. How much do six 87-pound sea otters weigh? _____

5. How much do seven 260-pound Atlantic green turtles weigh? _____

6. 6 × 54 = _____

7. _____ = 4 × 250

8. 2 × 460 = _____

9. _____ = 3 × 320

208 two hundred eight

Lesson 9·2 Number Stories

Use the Adult Weights of North American Animals poster on *Math Journal 2,* pages 204 and 205. Make up multiplication and division number stories. Ask a partner to solve your number stories.

1. _____

   Answer: _____

2. _____

   Answer: _____

3. _____

   Answer: _____

Date _____ Time _____

LESSON 9·2 Math Boxes

1. Nicky has $806 in the bank. Andrew has $589. How much more money does Nicky have than Andrew?

   $ _____

2. Write the numbers.

   5 tens 9 ones

   50 + _____ Total: 59

   3 tens 8 ones

   _____ + _____ Total: _____

3. _____ hour = 30 minutes

   _____ hours = 90 minutes

   2 hours = _____ minutes

   $1\frac{1}{4}$ hours = _____ minutes

   _____ hours = 180 minutes

4. Draw a shape with a perimeter of 14 units.

   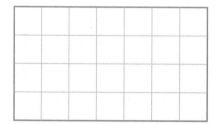

   What is the area of the shape?

   _____ square units

5. Fill in the missing numbers. Use fractions.

6. Circle the most appropriate unit.

   length of a calculator:

      inches   feet   miles

   weight of an adult:

      ounces   pounds   tons

   amount of gas in a car:

      cups   pints   gallons

210 two hundred ten

LESSON 9·3

Array Multiplication 1

1. How many squares are in a 4-by-28 array? Make a picture of the array.

   Total squares: _____

   4 × 28 = _____

2. How many squares are in a 3-by-26 array? Make a picture of the array.

   Total squares: _____

   3 × 26 = _____

3. How many squares are in a 6-by-32 array? Make a picture of the array.

   Total squares: _____

   6 × 32 = _____

two hundred eleven

LESSON 9·3 Geoboard Areas

Record your results in this table.

Geoboard Areas		
Area	Longer Sides	Shorter Sides
12 square units	units	units
12 square units	units	units
6 square units	units	units
6 square units	units	unit
16 square units	units	units
16 square units	units	units

1. Study your table. Can you find a pattern? _____

   _____

   _____

2. Find the lengths of the sides of a rectangle or square whose area is 30 square units without using the geoboard or geoboard dot paper. Make or draw the shape to check your answer. _____

3. Make check marks in your table next to the rectangles and squares whose perimeters are 14 units and 16 units.

212 two hundred twelve

Date _____ Time _____

LESSON 9·3 Math Boxes

1. If I wanted to have an equal chance of taking out a circle or a square, I would add _____ circle(s) to the bag.

2. Which number is the smallest? Fill in the circle for the best answer.

   Ⓐ 1,060

   Ⓑ 1,600

   Ⓒ 1,006

   Ⓓ 6,001

3. Write 3 fractions that are equivalent to $\frac{8}{12}$.

   _____ _____ _____

4. Pencils cost $1.99 for a package of 24 and $1.69 for a package of 16. What is the total cost of two 24-pencil packages and one 16-pencil package?

   Ballpark estimate: _____

   Exact answer: _____

5. Use bills and coins.

   Share $108 equally among 4 people.

   Each gets $_____.

   Share $61 equally among 4 people.

   Each gets $_____.

6. Measure the line segment to the nearest $\frac{1}{4}$ inch.

   _____

   about _____ in.

   Draw a line segment that is $2\frac{1}{4}$ inches long.

two hundred thirteen **213**

| Date | Time |

LESSON 9·4 Using the Partial-Products Algorithm

Multiply. Compare your answers with a partner. If you disagree, discuss your strategies with each other. Then try the problem again.

Example 7 × 46

$$\begin{array}{r} 46 \\ \times\ 7 \\ \hline \end{array}$$

7 [40s] → 280
7 [6s] → + 42
280 + 42 → 322

1. 34 × 2

$$\begin{array}{r} 34 \\ \times\ 2 \\ \hline \end{array}$$

2. 83 × 5

$$\begin{array}{r} 83 \\ \times\ 5 \\ \hline \end{array}$$

3. 55 × 6

$$\begin{array}{r} 55 \\ \times\ 6 \\ \hline \end{array}$$

4. 214 × 7

$$\begin{array}{r} 214 \\ \times\ \ \ 7 \\ \hline \end{array}$$

5. 403 × 5

$$\begin{array}{r} 403 \\ \times\ \ \ 5 \\ \hline \end{array}$$

| Date | Time |

LESSON 9·4 Measures

Measure these drawings to the nearest $\frac{1}{2}$ inch and $\frac{1}{2}$ centimeter.

1.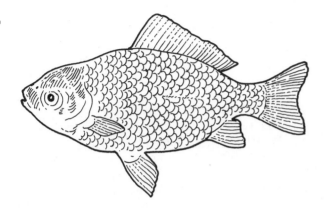

   The length of the fish

   about _____ in.   about _____ cm

2.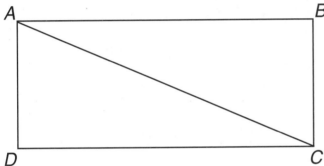

   The map distance from Alpha to Beta

   about _____ in.   about _____ cm

3. Line segment AB: about _____ in.

   Line segment AB: about _____ cm

   Line segment AC: about _____ in.

   Line segment AC: about _____ cm

Try This

Carefully draw the following line segments:

4. 9.5 cm

5. $4\frac{1}{4}$ in.

6. 2 cm shorter than 9.5 cm

7. $1\frac{1}{4}$ in. shorter than $4\frac{1}{4}$ in.

Date   Time

LESSON 9·4 Math Boxes

1. Morgan earned $252 shoveling snow. Casey earned $228. How much more money did Morgan earn? Fill in the circle for the best answer.

   Ⓐ $24   Ⓑ $26
   Ⓒ $470 Ⓓ $480

2. Write the numbers.

   5 hundreds 6 tens 4 ones

   _____ + _____ + _____

   Total: _____

   3 hundreds 2 tens 9 ones

   _____ + _____ + _____

   Total: _____

3. _____ seconds = 2 minutes

   28 days = _____ weeks

   6 months = _____ year

   _____ months = $1\frac{1}{2}$ years

4. The length of the longer side is _____ units.

   The length of the shorter side is _____ units.

   The area of the rectangle is _____ square units.

   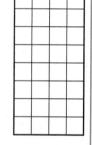

5. Fill in the missing numbers on the number line.

   1    $1\frac{3}{4}$

6. Circle the unit you would use to measure each item.

   weight of journal

   ounce   pound   ton

   length of football field

   inch   yard   mile

   length of paperclip

   cm   meter   kilometer

216 two hundred sixteen

LESSON 9·5

Shopping at the Stock-Up Sale

Use the Stock-Up Sale Poster #2 on page 217 in the *Student Reference Book*. Solve each number story below. There is no sales tax. Show how you got the answers.

1. When Mason sees bars of soap at the Stock-Up Sale, he wants to buy at least 5. He has $4.00. Can he buy 5 bars of soap? _____

   Number model: _____

   Can he buy 6 bars? _____

2. Vic's mom gave him a $5.00 bill to buy a toothbrush. If he goes to the sale, can he buy 5 toothbrushes? _____

   Exactly how much money does Vic need in order to be able to buy 5 toothbrushes at the sale price? _____

   Number model: _____

3. Andrea wants 2 bottles of glue. How much more will it cost her to buy 5 bottles at the sale price rather than 2 bottles at the regular price? _____

4. Make up a Stock-Up Sale story of your own.

   _____

   _____

   _____

   Answer: _____

   Number model: _____

two hundred seventeen **217**

Date _____ Time _____

LESSON 9·5 Math Boxes

1. Fill in the oval for the best answer. The perimeter of the quadrangle is

   ⊂⊃ 21 yd.

   ⊂⊃ 30 yd.   6 yd

   ⊂⊃ 24 yd.   15 yd

   ⊂⊃ 42 yd.

   SRB 150 151

2. Draw a 4-by-9 array of Xs.

   How many Xs in all? _____
   Write a number model.

   _____

   SRB 64 65

3. Use the partial-products algorithm to solve.

   9237
   $\underline{\times 60}\underline{\times 50}$

   SRB 68 69

4. Solve.

   $(40 \times 3) \div 2 =$ _____

   $4 \times (300 \div 6) =$ _____

   $(7 \times 80) + 140 =$ _____

   SRB 16

5. Draw a set of 12 circles.

   Color $\frac{5}{12}$ of the set blue.

   Color $\frac{1}{3}$ of the set red.

   Color $\frac{1}{6}$ of the set green.

   SRB 24

6. Solve.

   1 foot = _____ inches

   _____ feet = 36 inches

   1 yard = _____ feet

   _____ yards = 15 feet

   1 yard = _____ inches

   SRB 146

218 two hundred eighteen

LESSON 9·6

Factor Bingo Game Mat

Write any of the numbers 2 through 90 onto the grid above.

You may use a number only once.

To help you keep track of the numbers you use, circle them in the list.

```
         2    3    4    5    6    7    8    9   10
   11   12   13   14   15   16   17   18   19   20
   21   22   23   24   25   26   27   28   29   30
   31   32   33   34   35   36   37   38   39   40
   41   42   43   44   45   46   47   48   49   50
   51   52   53   54   55   56   57   58   59   60
   61   62   63   64   65   66   67   68   69   70
   71   72   73   74   75   76   77   78   79   80
   81   82   83   84   85   86   87   88   89   90
```

two hundred nineteen

Date _____ Time _____

LESSON 9·6 Using the Partial-Products Algorithm

Multiply. Show your work. Compare your answers with your partner's answers. If you disagree, discuss your strategies with each other. Then, try the problem again.

1.
       68
    × 2

2.
       96
    × 5

3.
       47
    × 4

4.
       85
    × 9

5.
      231
    ×   6

6.
      508
    ×   5

220 two hundred twenty

Date _____ Time _____

LESSON 9·6 Math Boxes

1. Estimate. Malachi sold 19 boxes of candy for $2.50 a box. About how much money should he have?

about _____

Number model:

_____

2. Solve.

$(9 \times 9) - (43 + 9) =$ _____

_____ $= (5{,}600 \div 80) \div 2$

_____ $= 963 + (567 - 439)$

3. Use your Fraction Cards. Write >, <, or = to make the number sentence true.

$\frac{1}{3}$ _____ $\frac{1}{4}$

$\frac{1}{3}$ _____ $\frac{4}{12}$

$\frac{1}{3}$ _____ $\frac{7}{8}$

$\frac{1}{3}$ _____ $\frac{4}{6}$

4. Use bills and coins.

Share $63.75 equally among 3 people.

Each gets $_____.

Share $63.00 equally among 5 people.

Each gets $_____.

5. You and a friend are playing a game with a 6-sided die. You win if an odd number is rolled. Your friend wins if an even number is rolled. Do you think this game is fair? Circle one.

yes   no

6. Measure this line segment.

_____

It is about _____ inches long.

It is about _____ centimeters long.

LESSON 9·7 Sharing Money

Work with a partner. Put your play money in a bank for both of you to use.

1. If $54 is shared equally by 3 people, how much does each person get?

   a. How many $10 bills does each person get? _____ $10 bill(s)

   b. How many dollars are left to share? $_____

   c. How many $1 bills does each person get? _____ $1 bill(s)

   d. Number model: $54 ÷ 3 = $_____

2. If $204 is shared equally by 6 people, how much does each person get?

   a. How many $100 bills does each person get? _____ $100 bill(s)

   b. How many $10 bills does each person get? _____ $10 bill(s)

   c. How many dollars are left to share? $_____

   d. How many $1 bills does each person get? _____ $1 bill(s)

   e. Number model: $204 ÷ 6 = $_____

3. If $71 is shared equally by 5 people, how much does each person get?

   a. How many $10 bills does each person get? _____ $10 bill(s)

   b. How many dollars are left to share? $_____

   c. How many $1 bills does each person get? _____ $1 bill(s)

   d. How many $1 bills are left over? _____ $1 bill(s)

   e. If the leftover $1 bill(s) are shared equally, how many cents does each person get? $_____

   f. Number model: $71 ÷ 5 = $_____

4. $84 ÷ 3 = $_____

5. $75 ÷ 6 = $_____

6. $181 ÷ 4 = $_____

7. $617 ÷ 5 = $_____

Lesson 9·7 Math Boxes

1. Draw a shape with a perimeter of 20 centimeters.

   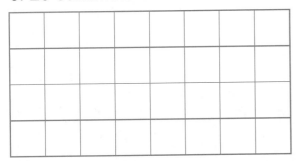

   What is the area of your shape?

   _____ square centimeters

2. Draw a 4-by-8 array of Xs.

   How many Xs in all? _____
   Write a number model.

   _____

3. Use the partial-products algorithm to solve.

   ```
     296        183
   ×   4      ×   7
   ```

4. Put in the parentheses needed to complete the number sentences.

   $15 + 80 \times 90 = 7{,}215$

   $14 - 6 \times 800 = 6{,}400$

   $60 \times 79 + 1 = 4{,}800$

5. What part of this pizza has been eaten?

   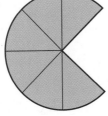

   _____

   What part is left?

   _____

6. Complete.

   24 inches = _____ feet

   30 cm = _____ mm

   _____ yards = 12 feet

   _____ yards = 72 inches

   4 meters = _____ centimeters

two hundred twenty-three **223**

Date _____ Time _____

LESSON 9·8 Division with Remainders

Solve the problems below. Remember that you will have to decide
what the remainder means in order to answer the questions.
You may use your calculator, counters, play money, or pictures.

1. Ruth is buying soda for a party. There are
   6 cans in a pack. She needs 44 cans.
   How many 6-packs will she buy?   _____ 6-packs

2. Paul is buying tickets to the circus.
   Each ticket costs $7. He has $47.
   How many tickets can he buy?   _____ tickets

3. Héctor is standing in line for the roller coaster.
   There are 33 people in line.
   Each roller coaster car holds 4 people.
   How many cars are needed to hold 33 people? _____ cars

Pretend that the division key on your calculator is broken.
Solve the following problems:

4. Regina is building a fence around her dollhouse.
   She is making each fence post 5 inches tall.
   The wood she bought is 36 inches long.
   How many fence posts does each piece of wood make? _____ posts

   Explain how you found your answer.

   _____

   _____

5. Missy, Ann, and Herman found a $10 bill.
   They want to share the money equally.
   How much money will each person get? _____

   Explain how you found your answer.

   _____

   _____

224 two hundred twenty-four

Date _____ **Time** _____

LESSON 9·8 Math Boxes

1. Kevin and Naomi read to each other for 35 minutes each day. About how many hours do they read to each other in one week?

   Answer: about _____
   (unit)

2. Fill in the circle for the best answer. $5 \times (6 - 5) =$ _____

   Ⓐ 4

   Ⓑ 5

   Ⓒ 16

   Ⓓ 25

3. Write 5 names for $\frac{3}{4}$.

4. Use bills and coins.

   Share $78 equally among 3 people.

   Each person gets $____.____.

   Share $53 equally among 4 people.

   Each person gets $____.____.

5. You and a friend are playing a game with the spinner. You win if the spinner lands on purple. Your friend wins if the spinner lands on black. Do you think this game is fair?

   yes

   no

   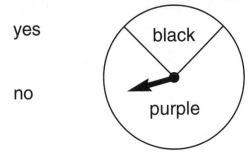

6. Draw a line segment $1\frac{3}{4}$ inches long.

   Draw a line segment $\frac{1}{2}$ inch longer than the one you just drew.

two hundred twenty-five **225**

Lesson 9·9: Lattice Multiplication

Megan has a special way of doing multiplication problems. She calls it lattice multiplication. Can you figure out how she does it?

Study the problems and solutions in Column A. Then try to use lattice multiplication to solve the problems in Column B.

Column A

3 × 64 = __192__

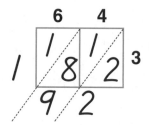

5 × 713 = __3,565__

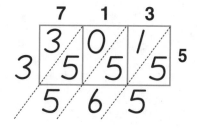

7 × 376 = __2,632__

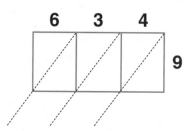

Column B

1. 4 × 65 = _____

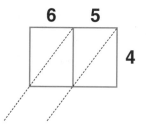

2. 6 × 815 = _____

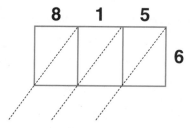

3. 9 × 634 = _____

226   two hundred twenty-six

Lesson 9·9 Lattice Multiplication Practice

1. 8 × 45 = _____

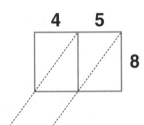

2. 9 × 37 = _____

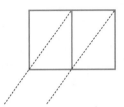

3. 5 × 23 = _____

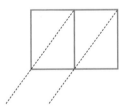

4. 3 × 124 = _____

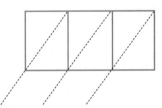

5. 6 × 431 = _____

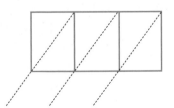

6. 7 × 209 = _____

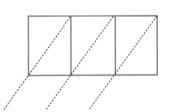

two hundred twenty-seven **227**

Date _____ Time _____

LESSON 9·9 Math Boxes

1. Name the eight factors of 24.

   ____, ____, ____, ____,

   ____, ____, ____, ____

2. Use the partial-products algorithm to solve. Show your work.

   $$238 \times 6 \qquad 574 \times 5$$

3. 16 books in all. 3 books per shelf.

   How many shelves? _____

   How many books left over? _____

4. Draw an angle that measures between 0° and 90°.

5. This shape is a _____.

   It has ____ sides and ____ vertices.

6. What is the median number of hours children sleep each night?

   ____ hours

Hours	Number of Children
8	////
9	//// ////
10	////
11	/

228 two hundred twenty-eight

Date	Time

LESSON 9·10 Array Multiplication 2

1. How many squares are in a 20-by-13 array?

Total squares = _____

20 × 13 = _____

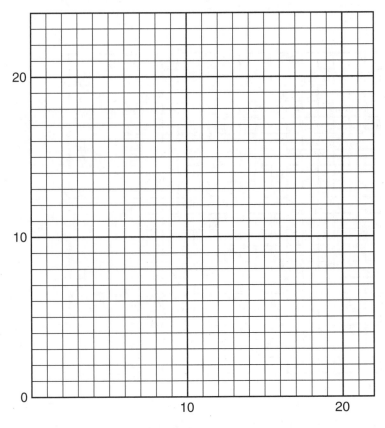

2. How many squares are in an 18-by-30 array?

Total squares = _____

18 × 30 = _____

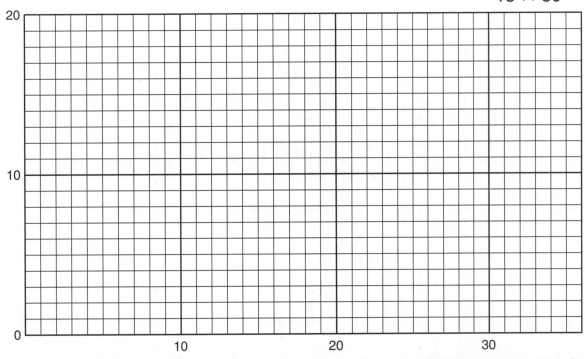

two hundred twenty-nine **229**

Date _____ Time _____

LESSON 9·10 Array Multiplication 3

1. How many squares are in a 17-by-34 array?   Total squares = _____

   $17 \times 34 =$ _____

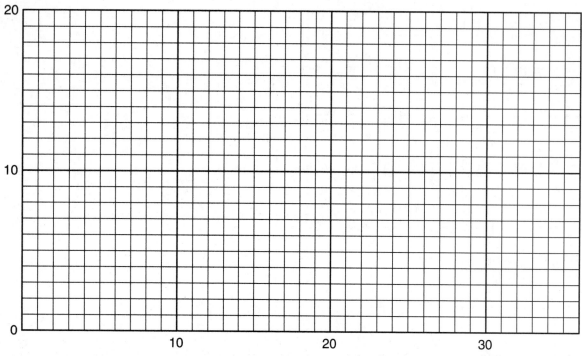

2. How many squares are in a 22-by-28 array?   Total squares = _____

   $22 \times 28 =$ _____

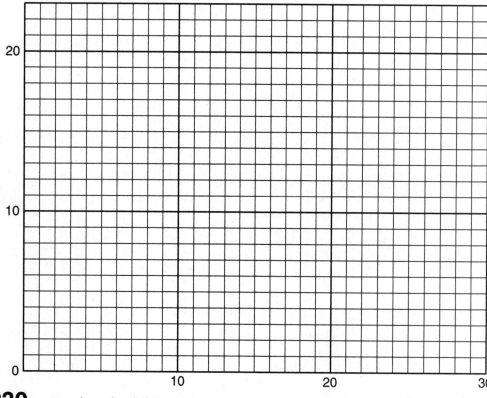

LESSON 9·10 Sharing Money

Work with a partner.

Materials
- ☐ number cards 0–9 (at least 2 of each)
- ☐ 1 die
- ☐ $10 bills, $1 bills, and tool-kit coins (optional)

Draw 2 number cards. Form a 2-digit number to show how much money will be shared. Roll 1 die to show how many friends will share the money. Fill in the boxes below.

1. $_____ is shared equally by _____ friends.

   a. How many $10 bills does each friend get? _____
   b. How many $1 bills does each friend get? _____
   c. How many $1 bills are left over? _____
   d. If the leftover money is shared equally, how many cents does each friend get? _____
   e. Each friend gets a total of $_____.
   f. Number model: _____

Repeat. Draw the next 2 cards. Roll the die. Fill in the blanks below.

2. $_____ is shared equally by _____ friends.

   a. How many $10 bills does each friend get? _____
   b. How many $1 bills does each friend get? _____
   c. How many $1 bills are left over? _____
   d. If the leftover money is shared equally, how many cents does each friend get? _____
   e. Each friend gets a total of $_____.
   f. Number model: _____

two hundred thirty-one **231**

Date _____ Time _____

LESSON 9·10 Math Boxes

1. Draw a shape with an area of 10 square centimeters.

   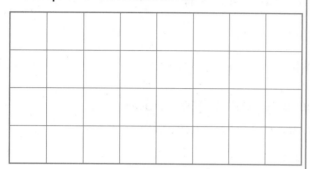

   What is the perimeter of your shape? _____ centimeters

2. Practice lattice multiplication.

   39 × 48 = _____

   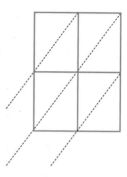

3. Make an estimate. About how much money, without tax, will you need for 5 gallons of milk that cost $3.09 each?

   about _____

4. Estimate. Reuben buys 3 bags of baby carrots at $2.19 per bag. He gives the cashier a $10 bill. About how much change should he get?

   about _____

5. Fill in the circle for the best answer. The turn of the angle is

   ○ A. greater than a $\frac{1}{4}$ turn.
   ○ B. less than a $\frac{1}{4}$ turn.
   ○ C. greater than a $\frac{1}{2}$ turn.
   ○ D. a full turn.

6. Circle the tool you would use to find the length of a pen:

   ruler   compass   scale

   the weight of a dime:

   ruler   compass   scale

   the way to get home:

   ruler   compass   scale

232   two hundred thirty-two

LESSON 9·11 Multiplication with Multiples of 10

Multiply. Compare your answers with a partner. If you disagree, discuss your strategies with each other. Then try the problem again.

Example:

```
    30
  × 26
```

20 [30s] → 600
6 [30s] → +180
             780

1.
```
    70
  × 18
```

2.
```
    88
  × 40
```

3.
```
    60
  × 35
```

4.
```
    80
  × 44
```

5.
```
    90
  × 63
```

two hundred thirty-three **233**

Date _____ Time _____

LESSON 9·11 Math Boxes

1. Write the six factors of 20.

   ____, ____, ____,

   ____, ____, ____

2. Use the partial-products algorithm to solve.

   ```
     489        608
   ×   7      ×   9
   ```

3. Allison has 58 stickers. She wants to share them among 8 friends.

   How many stickers does each friend get?

   _____

   How many stickers are left over?

   _____

4. Fill in the oval for the best answer. The degree measure of the angle is

   ○ less than 40°.

   ○ more than 100°.

   ○ more than 180°.

   ○ 90°.

5. Fill in the oval for the best answer. This picture of a 3-dimensional shape is called a

   ○ rectangular prism.

   ○ pyramid.

   ○ sphere.

   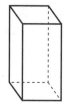

   It has ____ faces.

6. Number of children per classroom:

   25, 30, 26, 28, 33, 35, 28

   Median: _____

   Maximum: _____

   Minimum: _____

   Range: _____

Date _____ Time _____

LESSON 9·12 2-Digit Multiplication

Multiply. Compare your answers with a partner. If you disagree, discuss your strategies with each other. Try the problem again.

1.   24
   × 16

2.   42
   × 31

3.   12
   × 87

4.   59
   × 79

5.   36
   × 14

6.   42
   × 53

7. Describe in words how you solved Problem 1.

_____
_____
_____
_____

two hundred thirty-five **235**

Date _____ Time _____

LESSON 9·12 Math Boxes

1. Find the area of the rectangle.

   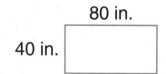
   80 in.
   40 in.

   ____ × ____ = ____ in.²
   length of   length of   area
   short side long side

2. Practice lattice multiplication.

   56 × 78 = _____

   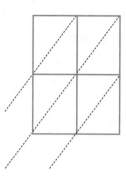

3. Taylor is 9 years old. About how many *days* old is he? Show your work.

   about _____ (unit)

   Number model: _____

4. Darius took two $5 bills and four $1 bills to the store. He bought shoelaces for $1.27 and 2 packs of batteries for $3.59 each. What is the smallest amount of money he can give to the cashier?

   _____

   _____

5. Fill in the oval for the best answer. The degree measure of the angle is

   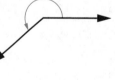

   ⊙ less than 90°.

   ⊙ less than 180°.

   ⊙ more than 180°.

6. Match the tool with its use.

   find weight   ruler

   measure length   clock

   tell time   scale

236   two hundred thirty-six

LESSON 9·13 — **Number Stories with Positive & Negative Numbers**

Solve the following problems. Use the thermometer scale, the class number line, or other tools to help.

1. The largest change in temperature in a single day took place in January 1916 in Browning, Montana. The temperature dropped 100°F that day. The temperature was 44°F when it started dropping.

   How low did it go? _____

2. The largest temperature rise in 12 hours took place in Granville, North Dakota, on February 21, 1918. The temperature rose 83°F that day. The high temperature was 50°F.

   What was the low temperature? _____

3. On January 12, 1911, the temperature in Rapid City, South Dakota, fell from 49°F at 6 A.M. to −13°F at 8 A.M.

   By how many degrees did the temperature drop in those 2 hours? _____

4. The highest temperature ever recorded in Verkhoyansk, Siberia, was 98°F. The lowest temperature ever recorded there was −94°F.

   What is the difference between those two temperatures? _____

5. Write your own number story using positive and negative numbers.

_____

_____

_____

two hundred thirty-seven **237**

LESSON 9·13 Math Boxes

1. length = _____ units

   width = _____ units

   area = _____ square units

   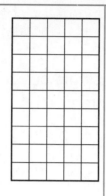

   2 factors of 45 are _____ and _____.

2. Use the partial-products algorithm to solve.

   $$\begin{array}{r} 652 \\ \times3 \\ \hline \end{array} \qquad \begin{array}{r} 408 \\ \times8 \\ \hline \end{array}$$

3. There are 347 candles. A box holds 50 candles. How many full boxes of candles is that?

   _____ boxes

   How many candles are left over?

   _____ candles

4. Fill in the oval for the best answer. The degree measure of the angle is

   ○ 180°.

   ○ less than 90°.

   ○ less than 270°.

   ○ more than 270°.

5. What 3-D shape is this a picture of? Fill in the oval for the best answer.

   ○ sphere

   ○ cylinder

   ○ pyramid

   What is the shape of the base?

   _____

6. Number of pets children have:

   0, 4, 0, 1, 1, 3, 6, 2, 5

   Median: _____

   Maximum: _____

   Minimum: _____

   Range: _____

238 two hundred thirty-eight

LESSON 9·14 Math Boxes

1. Measure this line segment to the nearest $\frac{1}{8}$ inch.

   _____

   It is about _____ inches long.

   Draw a line segment $1\frac{3}{4}$ inches long.

2. Measure this line segment to the nearest $\frac{1}{2}$ centimeter.

   _____

   It is about ____ centimeters long.

   Draw a line segment 3.5 centimeters long.

3. Circle the most appropriate unit.

   length of calculator

   inches   feet   miles

   weight of a third grader

   ounces   pounds   tons

   amount of water in a drinking glass

   cups   quarts   gallons

4. Find the median of the following numbers.

   34, 56, 34, 16, 33, 27, 45

   Median: ____

5. Find the maximum, minimum, and range of the following numbers:

   18, 13, 6, 9, 15, 25, 21, 17

   Maximum: _____

   Minimum: _____

   Range: _____

6. Circle the tool you use to find

   the temperature:

   scale   thermometer   ruler

   the weight of a *Student Reference Book*:

   scale   thermometer   ruler

   the perimeter of a *Student Reference Book*:

   scale   thermometer   ruler

two hundred thirty-nine 239

Date	Time

LESSON 10·1 Review: Units of Measure

1. Measure in centimeters. Which is longer, the path from A to B or the path from C to D? _____

   How much longer is it? _____

2. On the top edge of the ruler, make a dot at $3\frac{1}{2}$ inches. Label it E.

3. Make a dot at $4\frac{3}{4}$ in. Label it F.

4. Make a dot at $2\frac{7}{8}$ in. Label it G.

5. What is the distance from E to F? _____ in.

6. From E to G? _____ in.

7. From F to G? _____ in.

Complete.

8. 3 yd = _____ ft

9. 4 yd 1 ft = _____ ft

10. 1 ft 8 in. = _____ in.

11. 7 ft = _____ yd _____ ft

Measure the sides of the rectangle in centimeters. Find the area.

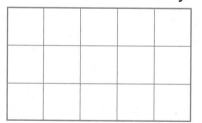

12. Area: _____
    (unit)

Try This

Measure the sides of the rectangle in centimeters. Find the area.

13. Area: _____
    (unit)

240 two hundred forty

Date _____ Time _____

LESSON 10·1   Weight and Volume

Complete Part 1 before the start of Lesson 10-3.

Part 1 Order the objects on display from heaviest (1) to lightest (4). Lift them to help you guess. Record your guesses below.

1. _____   2. _____

3. _____   4. _____

Complete Part 2 as part of Lesson 10-3.

Part 2 Record the actual order of the objects from heaviest (1) to lightest (4). Were your guesses correct?

1. _____   2. _____

3. _____   4. _____

Complete Parts 3 and 4 as part of Lesson 10-4.

Part 3 Order the objects on display from largest (1) to smallest (4) volume. Record your guesses below.

1. _____   2. _____

3. _____   4. _____

Part 4 Record the actual order of the objects from largest (1) to smallest (4) volume. Were your guesses correct?

1. _____   2. _____

3. _____   4. _____

1. Make a dot on the produce scale to show 2 lb. Label it *A*.

2. Make a dot on the produce scale to show $3\frac{1}{2}$ lb. Label it *B*.

3. Make a dot on the produce scale to show 2 lb, 8 oz. Label it *C*.

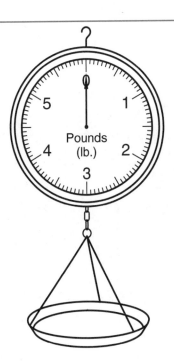

two hundred forty-one   **241**

Date _____ Time _____

Lesson 10·1 Multiplication Practice

Use your favorite multiplication algorithm to solve the following problems. Show your work.

1. 427
    ×   3

2. 505
    ×   8

3. 20
    × 90

4. 67
    × 40

5. 74
    × 35

6. 37
    × 58

242   two hundred forty-two

Date _____ Time _____

LESSON 10·1 Math Boxes

1. Measure the line segment to the nearest $\frac{1}{2}$ centimeter.

_____

_____

2. Circle the units you would use to measure each item.

height of a third grader
    inches   miles   square yards

length of a basketball court
    miles   feet   inches

3. Fill in the circle next to the picture of the triangular prism.

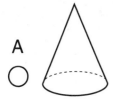 A ○
 B ○
 C ○
 D ○

4. Rectangle HFCD is a(n)

___ -by- ___ rectangle.

The area of rectangle HFCD:

___ × ___ = ___ square units.

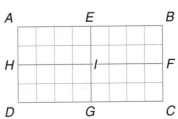

5. Practice lattice multiplication.

84 × 56 = _____

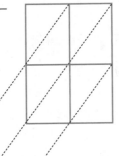

6. Points scored by players in a basketball game: 15, 22, 11, 12, 5

The maximum number of points is ____.

The minimum number of points is ____.

The range is ____ points.

two hundred forty-three   **243**

Volumes of Boxes

LESSON 10·2

Part 1 Use the patterns on *Math Masters,* page 323 to build Boxes A, B, C, and D. Record the results in the table.

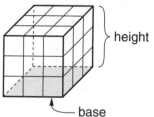

Box	Number of cm Cubes		Area of Base (square cm)	Height (cm)	Volume (cubic cm)
	Estimate	Exact			
A					
B					
C					
D					

Part 2 The following patterns are for Boxes E, F, and G. Each square stands for 1 square centimeter. Find the volume of each box. (Do not cut out the patterns.)

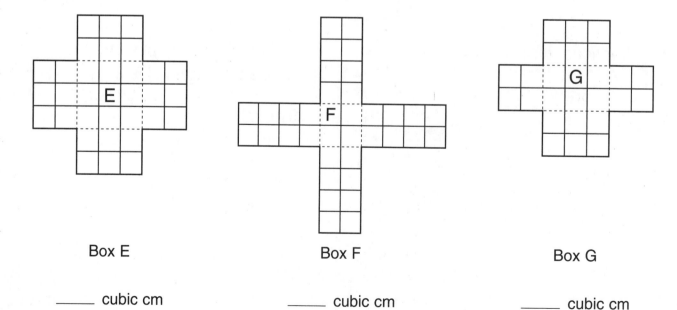

Box E _____ cubic cm

Box F _____ cubic cm

Box G _____ cubic cm

Lesson 10·2 Math Boxes

1. Circle any measurements in Column B that match the one in Column A.

Column A	Column B	
2 feet	12 in.	3 yd
	24 in.	1 yd
3 feet	36 in.	1 m
	1 yd	30 in.
2 yards	50 in.	72 in.
	6 ft	9 ft

2. Use the partial-products algorithm to solve.

   86   91
 × 27   × 64
 ─────   ─────

3. There are 20 crayons in a box. $\frac{1}{2}$ of the crayons are broken. How many crayons are broken?

   _____ crayons

   $\frac{1}{4}$ of the crayons are red.

   How many crayons are red?

   _____

4. Fill in the circle next to the numbers that are in order from smallest to largest.

   Ⓐ 0, 6, −3, 0.15

   Ⓑ 6, 0.15, 0, −3

   Ⓒ 0.15, 0, −3, 6

   Ⓓ −3, 0, 0.15, 6

5. Write the number that has
   2 in the thousandths place
   6 in the ones place
   3 in the hundredths place
   4 in the tenths place

   ___ . ___ ___ ___

6. Complete the "What's My Rule?" table.

   in ↓
   Rule
   Add 25 minutes
   ↓ out

in	out
7:00	
3:15	
5:45	
	7:40
	11:10

245

LESSON 10·3

Various Scales

Refer to pages 165 and 166 in your *Student Reference Book*. For each scale shown, list three things you could weigh on the scale.

balance scale

_____

_____

_____

market scale

_____

_____

_____

package scale

_____

_____

_____

bath scale

_____

_____

_____

spring scale

_____

_____

_____

produce scale

_____

_____

_____

letter scale

_____

_____

_____

platform scale

_____

_____

_____

infant scale

_____

_____

_____

diet/food scale

_____

_____

_____

| Date | Time |

Lesson 10·3 — Reading Scales

Read each scale and record the weight.

1. _____ lb

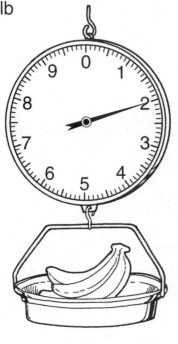

2. _____ oz

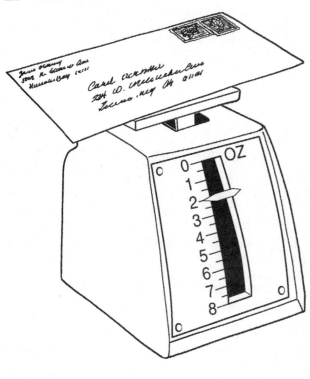

3. _____ lb

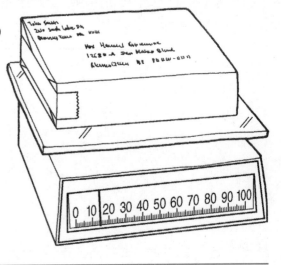

4. _____ lb

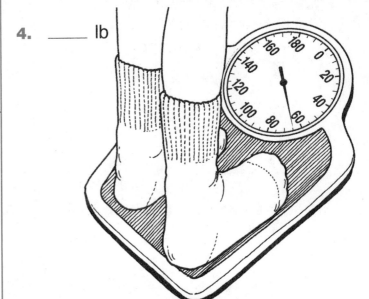

5. _____ g

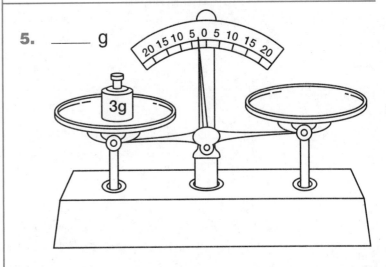

two hundred forty-seven **247**

Date _____ Time _____

LESSON 10·3 Math Boxes

1. Measure the line segment to the nearest $\frac{1}{2}$ inch.

   _____

   _____

2. Circle the units you would use to measure each item.

   length of a swimming pool
       meters   kilometers   centimeters

   length of an ant
       meters   kilometers   millimeters

3. This is a picture of a 3-dimensional shape. Name the shape.

   _____

   How many vertices does it have?

   _____

4. The length of the longer side is ___ units.

   The length of the shorter side is ___ units.

   The area of the rectangle is ___ square units.

   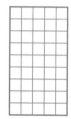

5. Practice lattice multiplication.

   75 × 64 = _____

   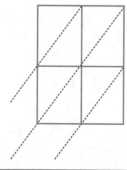

6. Laps completed during practice by members of the swim team:
   10, 15, 20, 15, 15

   The median number of laps completed is _____.

   The mode number of laps is _____.

248   two hundred forty-eight

Date _____ Time _____

LESSON 10·4 Math Boxes

1. Write equivalent lengths.

   $\frac{1}{3}$ yd = _____ ft

   18 in. = _____ yd

   50 mm = _____ cm

   0.6 m = _____ cm

2. Use the partial-products algorithm to solve. Show your work.

   $\quad\;\;36 \qquad\qquad 43$
   $\underline{\times 25} \qquad\quad \underline{\times 65}$

3. Complete the fraction number story.

   Samantha ate $\frac{\square}{8}$ of the pizza.

   Luke ate $\frac{\square}{8}$ of the pizza.

   Connor ate $\frac{\square}{8}$ of the pizza.

   $\frac{\square}{8}$ of the pizza was left over.

4. Find the distance between each pair of numbers.

   2 and −6 _____

   −7 and 15 _____

   100 and −500 _____

5. In the number 42.368:

   the 3 means ___3 tenths___

   the 2 means _____

   the 8 means _____

   the 6 means _____

   the 4 means _____

6. Tatiana gets her teeth cleaned every 6 months. If her last appointment was in February, when is her next appointment?

   _____

two hundred forty-nine **249**

Date Time

LESSON 10·5 Units of Measure

Mark the unit you would use to measure each item.

1. thickness of a dime ◯ millimeter ◯ gram ◯ foot
2. flour used in cooking ◯ gallon ◯ cup ◯ liter
3. gasoline for a car ◯ fluid ounce ◯ ton ◯ gallon
4. distance to the moon ◯ foot ◯ square mile ◯ kilometer
5. area of a floor ◯ square foot ◯ cubic foot ◯ foot
6. draperies ◯ kilometer ◯ millimeter ◯ yard
7. diameter of a basketball ◯ mile ◯ inch ◯ square inch
8. perimeter of a garden ◯ yard ◯ square yard ◯ centimeter
9. spices in a recipe ◯ teaspoon ◯ pound ◯ fluid ounce
10. weight of a nickel ◯ pound ◯ gram ◯ inch
11. volume of a suitcase ◯ square inch ◯ foot ◯ cubic inch
12. length of a cat's tail ◯ centimeter ◯ meter ◯ yard

Mark the best answer.

13. How much can an 8-year-old grow in a year?

    ◯ about 2 in.   ◯ about 2 ft   ◯ about 1 yd   ◯ about 1 m

14. How long would it take you to walk 3 miles?

    ◯ about 10 min   ◯ about 20 min   ◯ about 1 hour   ◯ about 5 hours

Date _____ Time _____

LESSON 10·5 Body Measures

Work with a partner to make each measurement to the nearest $\frac{1}{2}$ inch.

	Adult at Home	Me (Now)	Me (Later)
Date			
height	about _____ in.	about _____ in.	about _____ in.
shoe length	about _____ in.	about _____ in.	about _____ in.
around neck	about _____ in.	about _____ in.	about _____ in.
around wrist	about _____ in.	about _____ in.	about _____ in.
waist to floor	about _____ in.	about _____ in.	about _____ in.
forearm	about _____ in.	about _____ in.	about _____ in.
hand span	about _____ in.	about _____ in.	about _____ in.
arm span	about _____ in.	about _____ in.	about _____ in.
_____	about _____ in.	about _____ in.	about _____ in.
_____	about _____ in.	about _____ in.	about _____ in.

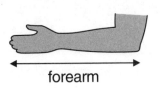
forearm

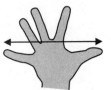
hand span

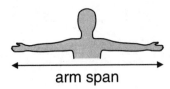
arm span

two hundred fifty-one 251

Lesson 10·5 Math Boxes

1. Solve. Show your work.

    654
  ×   7

2. Complete the bar graph.

Emma biked 4 miles.
Henry biked 5 miles.
Isaac biked 2 miles.

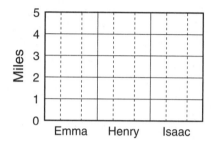

3. Circle the unit you would use to measure each item.

weight of journal

    ounce   pound   ton

length of car

    inch   yard   mile

length of paper clip

    centimeter   meter   kilometer

4. Cross out fractions less than $\frac{2}{3}$. Place a circle around the fractions equivalent to $\frac{2}{3}$.

$1\frac{2}{3}$ $\frac{1}{3}$

$\frac{4}{6}$ 2/5

$\frac{6}{9}$ $\frac{5}{6}$

5. Find the area of the rectangle.

60 cm
20 cm

____ cm × ____ cm = ____
length of short side   length of long side   area

6. Shade to show the following data.

A is 4 cm.

B is 3 cm.

C is 8 cm.

D is 7 cm.

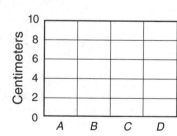

What is the range? _____

252   two hundred fifty-two

LESSON 10·6 A Mean, or Average, Number of Children

Activity 1 Make a bar graph of the data in the table.

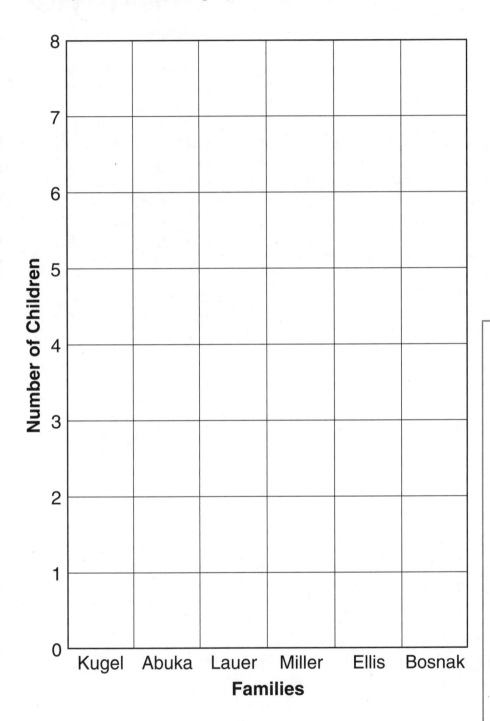

Family Sizes	
Family	Number of Children
Kugel	5
Abuka	1
Lauer	2
Miller	7
Ellis	1
Bosnak	2

Activity 2
(to be done later)

Use the table above. List the number of children in order.

_____

_____

_____

_____

_____

_____

The mean, or average, number of children in the six families in the table is _____.

The median number of children in the six families in the table is _____.

two hundred fifty-three **253**

LESSON 10·6 A Mean, or Average, Number of Eggs

Activity 1 Make a bar graph of the data in the table.

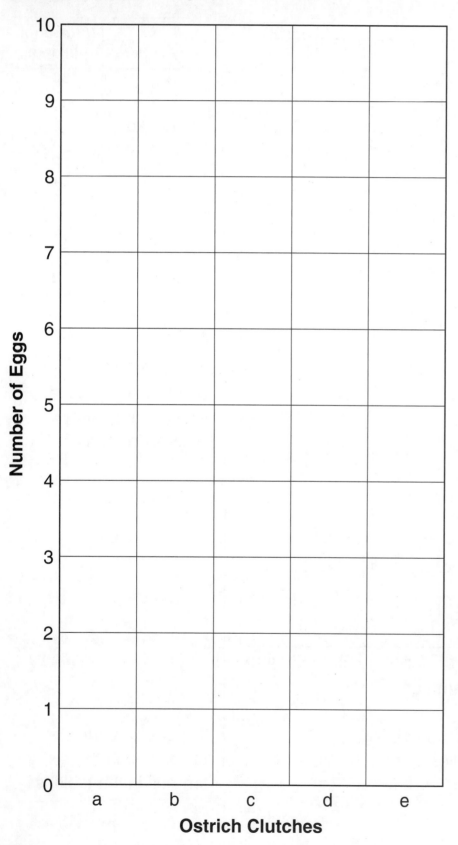

Ostrich Clutches

Clutch	Number of Eggs
a	6
b	10
c	4
d	2
e	8

The mean, or average, number of eggs in the five clutches is _____.

Activity 2
(to be done later)

List the number of eggs in the clutches in order.

_____

_____

_____

_____

_____

The median is _____ eggs per clutch.

254 two hundred fifty-four

Lesson 10·6 Math Boxes

1. Measure each side of the triangle to the nearest centimeter.

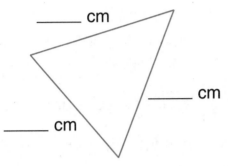

Perimeter = _____ cm

2. There are 5 blocks in a bag. 2 blocks are red, 2 blocks are blue, and 1 block is green. What are the chances of pulling out a red block?

_____ out of _____ chances

3. James built a rectangular prism out of base-10 blocks. He used 30 cm cubes to make the base. He put 4 more layers of cubes on top of that. What is the volume of the prism he built?

_____ cubic centimeters

4. Complete.

1 gallon = _____ quarts

_____ gallons = 12 quarts

1 pint = _____ cups

_____ pints = 14 cups

1 cup = _____ fl oz

_____ cups = 72 fl oz

5. Molly is playing with 5 toy cars. This is only $\frac{1}{3}$ of her set of cars. How many cars are in her complete set? Fill in the circle next to the best answer.

Ⓐ $\frac{5}{3}$ cars Ⓒ 10 cars

Ⓑ 5 cars Ⓓ 15 cars

6. Color the circle so that it matches the description.

$\frac{1}{2}$ blue

$\frac{1}{3}$ green

$\frac{1}{6}$ yellow

Which color would you expect the spinner to land on most often? _____

two hundred fifty-five **255**

LESSON 10·7 Finding the Median and the Mean

1. The median (middle) arm span in my class is about _____ inches.

2. The mean (average) arm span in my class is about _____ inches.

3. Look at page 251 in your journal. Use the measurements for an adult and the *second* measurements for yourself to find the median and mean arm spans and heights for your group. Record the results in the table below.

    a. Find the median and mean arm spans of the *adults* for your group.

    b. Find the median and mean arm spans of the *children* for your group.

    c. Find the median and mean heights of the *adults* for your group.

    d. Find the median and mean heights of the *children* for your group.

Summary of Measurements for Your Group

Measure	Adults	Children
Median arm span		
Mean arm span		
Median height		
Mean height		

Find the mean of each set of data. Use your calculator.

4. High temperatures: 56°F, 62°F, 74°F, 68°F   _____ °F

5. Low temperatures: 32°F, 42°F, 58°F, 60°F   _____ °F

6. Ticket sales: $710, $650, $905 $_____

7. Throws: 40 ft, 32 ft, 55 ft, 37 ft, 43 ft, 48 ft   _____ ft

Date _____ Time _____

Lesson 10·7 Math Boxes

1. Solve. Show your work.

       837
     ×   4
    _____

2. Read the graph. Which month had the most days of rain?

    _____

    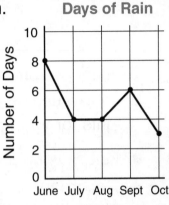
    Days of Rain

    What is the median number of days of rain? _____

3. Name 4 objects that weigh less than 1 pound.

    _____

    _____

    _____

    _____

4. Circle the fractions that are greater than $\frac{1}{4}$. Cross out the fractions that are equivalent to $\frac{1}{4}$.

    $\frac{2}{8}$ $\frac{4}{5}$

    $\frac{1}{2}$ $\frac{4}{12}$

    $\frac{3}{12}$ $\frac{2}{5}$

5. Fill in the oval next to the best answer. The area of the rectangle is

    ○ 9 sq units.

    ○ 14 sq units.

    ○ 18 sq units.

    ○ 140 sq units.

    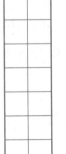

6.
    ```
    Number                    X
      of          X  X        X
    Children  X   X  X  X  X  X
              _____
              0   1  2  3  4  5
                Number of Fish
    ```

    The median is _____ fish.

two hundred fifty-seven   **257**

Date   Time

LESSON 10·8 Calculator Memory

For each problem:

- ◆ Clear the display and memory:
  Press [MRC] [MRC] [ON/C] or [AC].
- ◆ Enter the problem key sequence.
- ◆ Guess what number is in memory.
- ◆ Record your guess.
- ◆ Check your guess. Press [MRC] or [MR].
- ◆ Record the answer.

Your display should look like this

[0.]

before you start a new problem.

Problem Key Sequence	Your Guess	Number in Memory
1. 7 [M+] 9 [M+]		
2. 20 [M+] 16 [M−]		
3. 5 [M+] 10 [M+] 20 [M+]		
4. 12 [+] 8 [M+] 6 [M−]		
5. 15 [−] 9 [M+] 2 [M−]		
6. 25 [M+] 7 [+] 8 [M−]		
7. 30 [M+] 15 [−] 5 [M+]		
8. 2 [+] 2 [M+] 2 [−] 2 [M+] 2 [+] 2 [M+]		

9. Make up your own key sequences with memory keys. Ask a partner to enter the key sequence, guess what number is in memory, and then check the answer.

Problem Key Sequence	Your Guess	Number in Memory

Date _____ Time _____

LESSON 10·8 Measurement Number Stories

workspace

1. The gas tank of Mrs. Rone's car holds about 12 gallons. About how many gallons are in the tank when the gas gauge shows the tank to be $\frac{3}{4}$ full?

   _____

2. When the gas tank of Mrs. Rone's car is about half empty, she stops to fill the tank. If gas costs $1.25 per gallon, about how much does it cost to fill the tank?

   _____

Harry's room measures 11 feet by 13 feet. The door to his room is 3 feet wide. He wants to put a wooden border, or baseboard, around the base of the walls.

3. Draw a diagram of Harry's room on the grid below. Show where the door is. Let each side of a grid square equal 1 foot.

4. How many feet of baseboard must Harry buy? _____

5. How many yards is that? _____

6. If baseboard costs $4.00 a yard, how much will Harry pay? _____

workspace

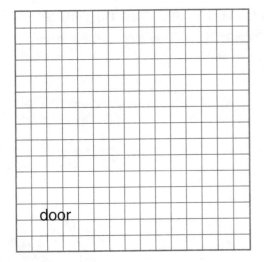

two hundred fifty-nine 259

Date _____ Time _____

Lesson 10·8 Math Boxes

1. Measure each side of the quadrangle to the nearest half-centimeter.

   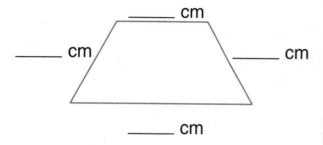

   Another name for this quadrangle is a _____.

2. Fill in the oval for the best answer. There are 6 blocks in a bag. 5 blocks are blue and 1 block is red. The chances of drawing the red block are:

   ⬭ 1 out of 6.
   ⬭ 5 out of 6.
   ⬭ 1 out of 5.
   ⬭ 5 out of 11.

3. Chanel built a rectangular prism out of base-10 blocks. She used 50 cm cubes to make the base. She put 9 more layers of cubes on top of that. What is the volume of the prism she built?

   _____ cubic centimeters

4. 1 quart = _____ pints

   _____ quarts = 16 pints

   1 quart = _____ fl oz

   _____ quarts = 96 fl oz

   1 gallon = _____ fl oz

5. There are 24 children in Mrs. Hiller's class. $\frac{1}{2}$ of the children play soccer. How many children play soccer?

   _____ children

   $\frac{1}{3}$ of the children play a musical instrument. How many children play a musical instrument?

   _____ children

6. Design a spinner that has an equal chance of landing on red or green.

   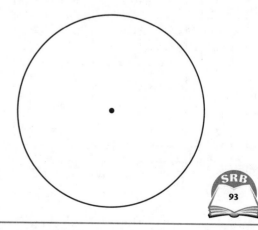

260 two hundred sixty

Date　　　　　　　　　　　　Time

LESSON 10·9　Frequency Table

1. Fill in the table of waist-to-floor measurements for the class. This kind of table is called a frequency table.

Waist-to-Floor Measurement (inches)	Frequency	
	Tallies	Number

Total =

2. What is the median (middle value) of the measurements? _____ in.

3. What is the mean (average) of the measurements? _____ in.

4. The *mode* is the measurement, or measurements, that occur most often. What is the mode of the waist-to-floor measurements for the class? _____ in.

LESSON 10·9 **Bar Graph**

Make a bar graph of the data in the frequency table on journal page 261.

Class Waist-to-Floor Measurements

Waist-to-Floor Measurements (in.)

Number of Children

Date _____ Time _____

LESSON 10·9 Math Boxes

1. Use the partial-products algorithm to solve.

   　83　　　　　72
   ×44　　　×36

2. 1 pint = _____ fluid ounces

   _____ pints = 48 fluid ounces

   1 half-gallon = _____ quarts

   _____ half-gallons = 6 quarts

   1 liter = _____ milliliters

3. Jerry has 16 fish in a tank. Color $\frac{3}{8}$ of the fish red, $\frac{1}{4}$ of the fish blue, and the rest yellow. What fraction of the fish are yellow?

   _____

4. Fill in the missing factors.

   40 × _____ = 280

   70 × _____ = 5,600

   8 × _____ = 24,000

   600 × _____ = 54,000

5. Weight in pounds of newborn babies: 11, 8, 8, 7, 6

   The mean (average) weight is

   ____ pounds.

   The median weight is ____ pounds.

6. On the first day of spring, the lengths of the day and night are equal. If the sun rises at 6:51 A.M. on that day, at what time would you expect it to set?

   ____ : ____ P.M.

two hundred sixty-three **263**

LESSON 10·10 Plotting Points on a Coordinate Grid

1. Draw a dot on the number line for each number your teacher dictates. Then write the number under the dot.

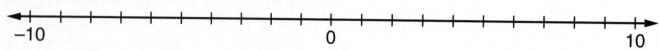

2. Draw a dot on the grid for each ordered pair. Write the letter for the ordered pair next to the dot.

   Sample A: (3,6)
   B: (3,4) C: (4,3) D: (1,2)
   E: (2,3) F: (5,2) G: (4,4)
   H: (4,0) I: (6,4) J: (0,5)
   K: (3,2) L: (5,4) M: (1,4)

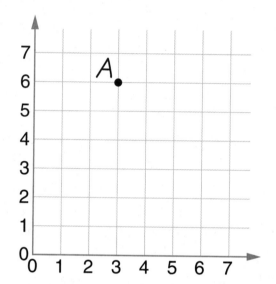

3. Do you know the answer to this riddle?

   Which two letters contain nothing? To find out, draw the following line segments on the grid: $\overline{MD}$, $\overline{ME}$, $\overline{EB}$, $\overline{BK}$, $\overline{GI}$, and $\overline{LF}$.

   _____

Draw the following line segments on the coordinate grid.

4. From (0,6) to (2,7); from (2,7) to (3,5); from (3,5) to (1,4); from (1,4) to (0,6)

   What kind of quadrangle is this?

   _____

5. From (7,0) to (7,4); from (7,4) to (5,3); from (5,3) to (5,1); from (5,1) to (7,0)

   What kind of quadrangle is this?

   _____

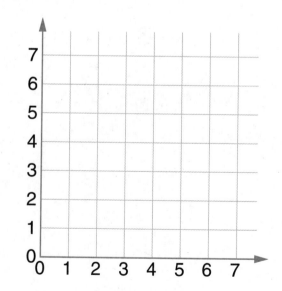

264 two hundred sixty-four

Date **Time**

Lesson 10·10 Math Boxes

1. Use the partial-products algorithm to solve.

```
    82         94
  × 35       × 76
```

2. There are 4 quarts in a gallon. How many quarts of paint did Sally use if she used $1\frac{1}{2}$ gallons of paint? Fill in the circle next to the best answer.

○ **A** 4 quarts

○ **B** 6 quarts

○ **C** 8 quarts

○ **D** 16 quarts

3. There are _____ books in $\frac{2}{5}$ of a set of 25 books.

There are _____ minutes in $\frac{3}{4}$ of an hour.

I have six books. This is $\frac{1}{6}$ of a set of books. How many books are in the complete set?

_____ books

4. 50 is 10 times as much as _____.

700 is _____ times as much as 7.

_____ is 100 times as much as 90.

60,000 is 1,000 times as much as _____.

5. Number of fish caught each weekend at Aunt Mary's lake:

3, 6, 5, 1, 7, 1, 5

The median number of fish caught: _____

The mean (average) number of fish caught: _____

6. Anchorage, Alaska has very long days in the summer. In the middle of July, the sun rises at about 3:20 A.M. and sets at about 10:20 P.M. About how many hours of daylight are there?

About _____ hours

two hundred sixty-five **265**

Date _____ Time _____

Lesson 10·11 — Math Boxes

1. What is the mode of the test scores for the class? _____ %

Test Score	Number of Children
100%	///
95%	//// /
90%	//// ///
85%	////

SRB 81

2. It is 7:45 A.M. Draw the hour and minute hands to show the time 15 minutes earlier. What time does the clock show now? _____

3.
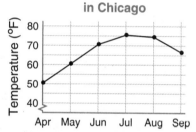

Average Monthly Temperature in Chicago

Which month has the highest average temperature?

_____

 SRB 90

4. Design a spinner that is as likely to land on blue as on yellow.

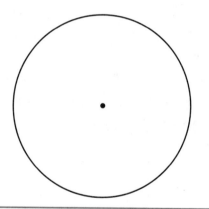

 SRB 93

5. 10 marbles in a jar. 100 random draws. You get:

♦ a black marble 32 times.

♦ a white marble 68 times.

How many marbles of each kind do you think are in the jar?

_____ black marbles

_____ white marbles

6. Record High Temperature for Five States

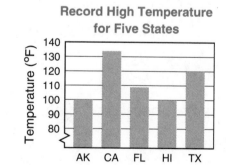

Which states have the same record high temperature?

_____

 SRB 86 87

Date _____ Time _____

LESSON 11·1 Math Boxes

1. Shade $\frac{3}{5}$ of the rectangle.

   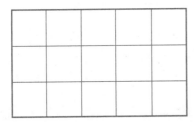

   What fraction is *not* shaded?

   _____

2. It is 8:05 A.M. Draw the hour and minute hands to show the time 15 minutes earlier. What time does the clock show?

   _____

3. Write the ordered pair for each letter on the grid.

   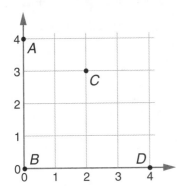

   A: (___ , ___)

   B: (___ , ___)

   C: (___ , ___)

   D: (___ , ___)

4. If I wanted to take out a square about 4 times as often as a circle, I would put in _____ square(s).

   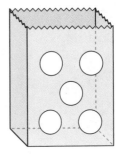

5. Write the number that is 100 **more** than 6,875,921.

   _____

   Write the number that is 1,000 **less** than 4,960,758.

   _____

   Read the numbers to a partner.

6. A large bag of candy costs $3.59. What is the cost of 6 bags?

   Fill in the oval next to the closest estimate.

   ◯ about $15.00

   ◯ about $18.00

   ◯ about $21.00

   ◯ about $27.00

two hundred sixty-seven **267**

Date _____ Time _____

Lesson 11·2 Math Boxes

1. Draw the hour and minute hands to show 11:22 A.M.

How long until 12:00 P.M.?

_____ hours _____ minutes

2. Use the partial-products algorithm to solve. Show your work.

   77   93
  × 24   × 61

3. It takes Linda and Craig 18 minutes to ride their bicycles to the library. If they leave home at 3:53 P.M., at what time will they arrive?

   _____ : _____ P.M.

4. A vase has 5 red flowers, 4 orange flowers, and 2 yellow flowers. If he doesn't look, what are the chances that Aaron will choose a red flower?

   _____ out of _____

5. What is the volume of the rectangular prism? Fill in the circle next to the best answer.

   Ⓐ 16 cubic units
   Ⓑ 32 cubic units
   Ⓒ 48 cubic units
   Ⓓ 64 cubic units

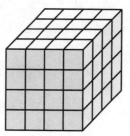

6. **Number of Sunny Days in Seattle**

How many sunny days were there in August? _____

268 two hundred sixty-eight

LESSON 11·3 **Spinners**

Math Message
Color each circle so that it matches the description.

1.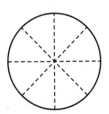

    $\frac{3}{4}$ red, $\frac{2}{8}$ blue

2.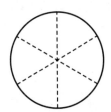

    $\frac{1}{2}$ red, $\frac{1}{3}$ yellow, $\frac{1}{6}$ blue

3.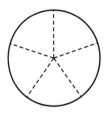

    $\frac{2}{5}$ red, $\frac{2}{5}$ blue, $\frac{1}{5}$ yellow

Spinner Experiments

Tape *Math Masters,* p. 367, to your desk or table. Make a spinner on the first circle.

4. Spin the paper clip 10 times. Tally the number of times the paper clip lands on the shaded part and on the white part.

5. Record results for the whole class.

Lands On	Tallies
shaded part	
white part	

Lands On	Totals
shaded part	
white part	

Make a spinner on the second circle.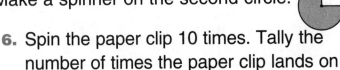

6. Spin the paper clip 10 times. Tally the number of times the paper clip lands on the shaded part and on the white part.

7. Record results for the whole class.

Lands On	Tallies
shaded part	
white part	

Lands On	Totals
shaded part	
white part	

8. With the second spinner, the paper clip has a better chance of landing on the _____ part of the spinner than on the _____ part.

two hundred sixty-nine

Date _____ Time _____

LESSON 11·3 Estimate, Then Calculate

For each problem, make a ballpark estimate and circle the phrase that best describes your estimate. Next, calculate the exact sum or difference. Check that your answer is close to your estimate.

1. more than 500

   less than 500

   ```
     825
   - 347
   ```

2. more than 500

   less than 500

   ```
     984
   - 392
   ```

3. more than 500

   less than 500

   ```
     658
   - 179
   ```

4. more than 500

   less than 500

   ```
     227
   + 285
   ```

5. more than 500

   less than 500

   ```
     324
   + 161
   ```

6. more than 500

   less than 500

   ```
     179
   + 338
   ```

270 two hundred seventy

Date　　　　　　　　　　　　Time

LESSON 11·3　Math Boxes

1. How many thirds are shaded?

   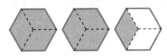

   _____ thirds

   Write the fraction: _____

   Write the mixed number: _____

2. Draw the hands to show 10:36.

   How many minutes until 11:16? _____

3. Write the ordered pair for each letter on the grid.

   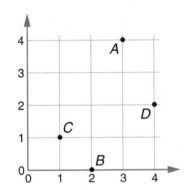

   A: (___, ___)

   B: (___, ___)

   C: (___, ___)

   D: (___, ___)

4. Design a spinner that is 3 times as likely to land on blue as it is to land on yellow.

   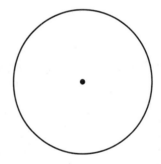

5. Write the number that is 10,000 **less** than 307,473.

   _____

   Write the number that is 100,000 **more** than 938,467.

   _____

   Read the numbers to a partner.

6. Fill in the oval next to the closest estimate.

   5,634 − 2,987 = _____

   ⚪ about 2,000

   ⚪ about 2,300

   ⚪ about 2,600

   ⚪ about 3,000

two hundred seventy-one　271

Lesson 11·4   Making Spinners

Math Message

1. Use exactly six different colors. Make a spinner so the paper clip has the **same chance** of landing on any one of the six colors.

   (*Hint:* Into how many equal parts should the circle be divided?)

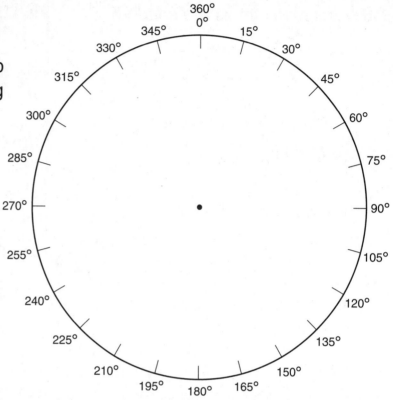

2. Use only blue and red. Make a spinner so the paper clip is **twice as likely** to land on blue as it is to land on red.

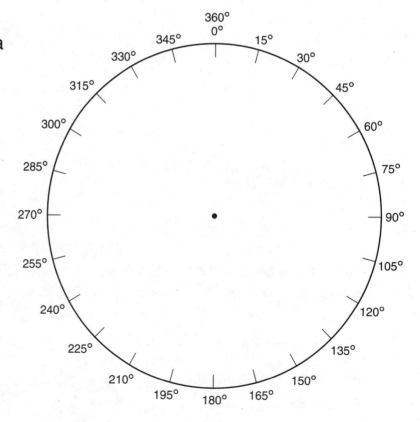

272   two hundred seventy-two

Lesson 11·4 Making Spinners continued

3. Use only blue, red, and green. Make a spinner so the paper clip:

   ◆ has the **same chance** of landing on blue and on red

   and

   ◆ is **less likely** to land on green than on blue.

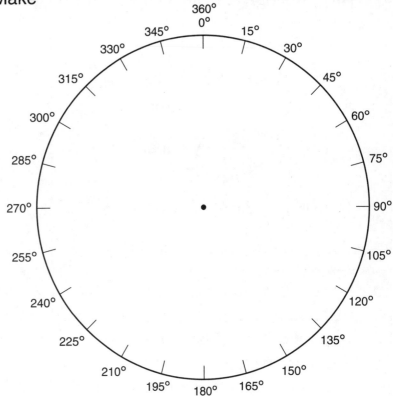

4. Use only blue, red, and yellow. Make a spinner so that the paper clip:

   ◆ is **more likely** to land on blue than on red

   and

   ◆ is **less likely** to land on yellow than on blue.

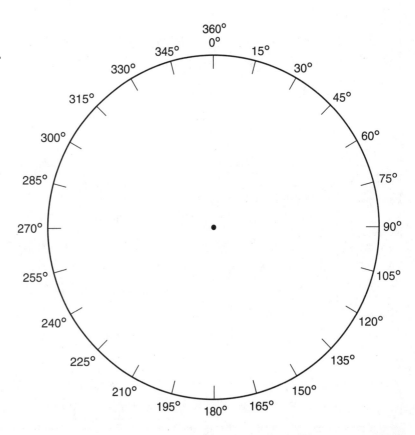

Lesson 11·4 Math Boxes

1. Draw the hands to show 9:34 A.M.

How long until 10:00 A.M.?

_____ hours _____ minutes

2. Solve. Show your work.

$$\begin{array}{r} 78 \\ \times\ 26 \\ \hline \end{array} \qquad \begin{array}{r} 56 \\ \times\ 92 \\ \hline \end{array}$$

3. It takes Cindy 20 minutes to take a bath, comb her hair, and brush her teeth. If she must be in bed by 8:00 P.M., what is the latest time she can start getting ready for bed?

_____ : _____ P.M.

4. What are the chances of pulling out a square block without looking?

_____ out of _____

5. Complete the table.

Area of Base (square feet)	Height (feet)	Volume (cubic feet)
40	90	
20	70	
800	9	
50	80	

6. Miles Run for Marathon Training

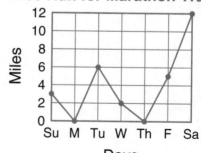

What is the median number of miles run this week? _____

Lesson 11·5

Random-Draw Problems

Each problem involves marbles in a jar. The marbles are blue, white, or striped. A marble is drawn at random (without looking) from the jar. The type of marble is tallied. Then the marble is returned to the jar.

♦ Read the description of the random draws in each problem.

♦ Circle the picture of the jar that best matches the description.

1. From 100 random draws, you get:

   a blue marble ● 62 times.

   a white marble ○ 38 times.

10 marbles in a jar

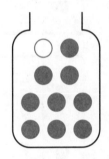
10 marbles in a jar

2. From 100 random draws, you get:

   a blue marble ● 23 times.

   a white marble ○ 53 times.

   a striped marble ⊘ 24 times.

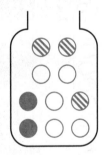
10 marbles in a jar

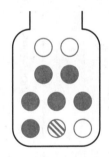
10 marbles in a jar

Try This

3. From 50 random draws, you get:

   a blue marble ● 30 times.

   a white marble ○ 16 times.

   a striped marble ⊘ 4 times.

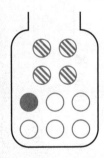
10 marbles in a jar

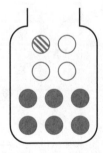
10 marbles in a jar

two hundred seventy-five

Date _____ Time _____

LESSON 11·5 Reading and Writing Numbers

Write the value of 7 for each column below.

L	K	,	J	,	I	H	G	,	F	E	D	.	C	B	A
hundred millions	ten millions	,	millions	,	hundred thousands	ten thousands	thousands	,	hundreds	tens	ones	.	tenths	hundredths	thousandths
7	7	,	7	,	7	7	7	,	7	7	7	.	7	7	7

Example: Column K: _70,000,000 or 70 millions_

1. Column A: _____
2. Column G: _____
3. Column F: _____
4. Column I: _____
5. Column C: _____
6. Column B: _____
7. Column L: _____

Write the numbers that your teacher dictates.

8. _____ 9. _____ 10. _____

11. _____ 12. _____ 13. _____

276 two hundred seventy-six

Date _____ Time _____

Lesson 11·5 Math Boxes

1. Jessica makes $3.25 every hour she works at the lemonade stand. Will she make enough money to buy a $12 watch if she works from 10:30 A.M. to 2:30 P.M.?

   _____

2. Number of books read during summer: 9, 9, 5, 15, 3, 9, 6

   The mean number of books read is _____.

   The median number of books read is _____.

3. Write the number that has

   3 in the hundred-thousands place,

   6 in the thousands place,

   4 in the ten-thousands place,

   1 in the millions place, and

   5 in all of the other places.

   ___ , ___ ___ ___ , ___ ___ ___

4. Color the spinner so it is more likely to land on blue than orange, and more likely to land on green than blue.

   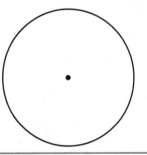

5. **2-Hour Water Polo Practice**

   Which two activities together make up $\frac{1}{2}$ of a 2-hour water polo practice?

   _____ and _____

6. Danielle skates from 6:45 to 7:30 every morning and from 3:05 to 3:55 in the afternoon on Mondays and Wednesdays. How long does she skate in a week?

   ____ hours ____ minutes

two hundred seventy-seven **277**

Lesson 11·6 Math Boxes

1. Fill in the oval next to the closest estimate.

    747 + 932 = _____

    ○ about 1,500

    ○ about 1,700

    ○ about 2,000

    ○ about 2,500

2. What is the median number of pets children have? _____ pet(s)

Number of Pets	Number of Children
0	///
1	//// ////
2	////
3	///
4	/
5	/

    What is the mode number of pets? _____ pet(s)

3. Write the number that has

    1 in the ten-thousands place,

    7 in the thousands place,

    2 in the hundred-thousands place,

    8 in the millions place, and

    0 in all of the other places.

    ___,___ ___ ___,___ ___ ___

4. Design a spinner that is twice as likely to land on blue as it is to land on yellow.

    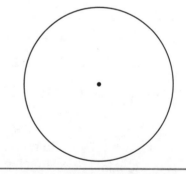

5. **Weights of 5 Dogs**

    (bar graph: Bernie 45, Gypsy 30, Butler 55, Lucky 40, Dusk 80; y-axis Pounds 0–80; x-axis Dogs)

    What is the range of weights?

    _____ (units)

6. Bria practices the piano from 2:45 P.M. to 3:25 P.M. every day after school and from 11:40 A.M. to 12:10 P.M. on weekends. How long does she practice the piano in one week?

    _____ hours _____ minutes

278 two hundred seventy-eight

Date | Time

LESSON 11·1 Sunrise and Sunset Record

Date	Time of Sunrise	Time of Sunset	Length of Day
			hr min
			hr min
			hr min
			hr min
			hr min
			hr min
			hr min
			hr min
			hr min
			hr min
			hr min
			hr min
			hr min
			hr min
			hr min
			hr min
			hr min
			hr min
			hr min
			hr min

LESSON 11·1 Length-of-Day Graph

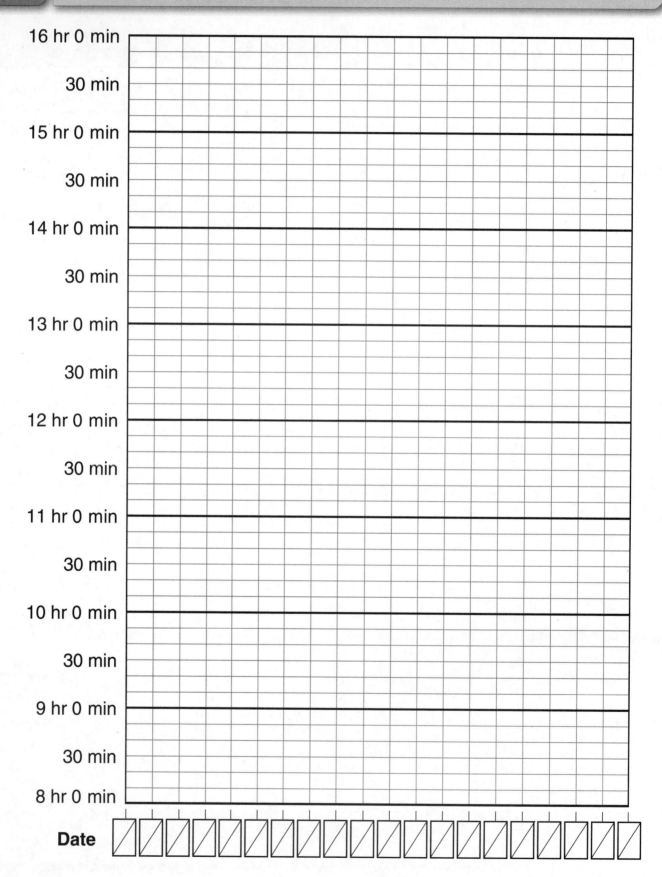

Date _____ Time _____

LESSON 11·1 **Length of Day Graph** *continued*

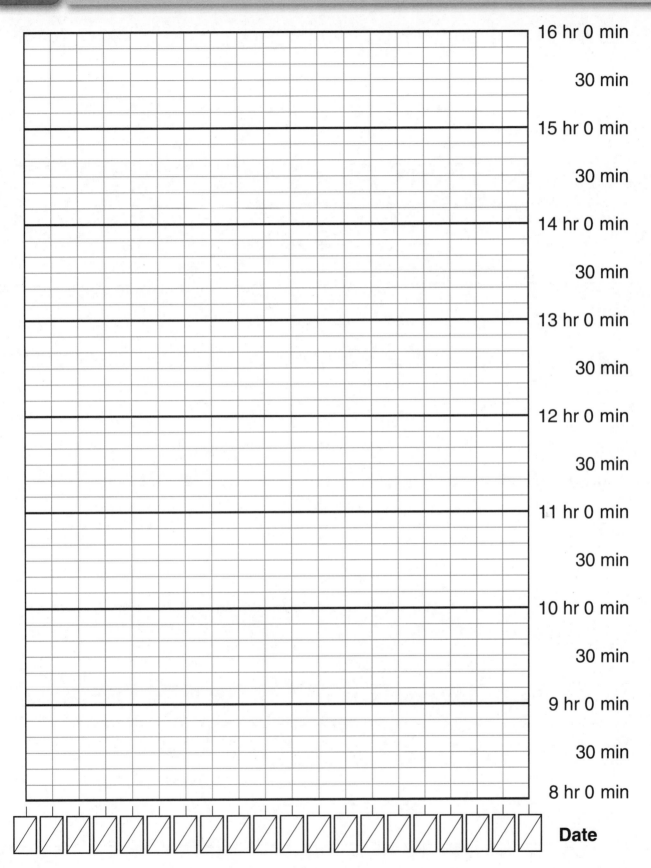

Date   Time

Notes

Date | Time

Notes

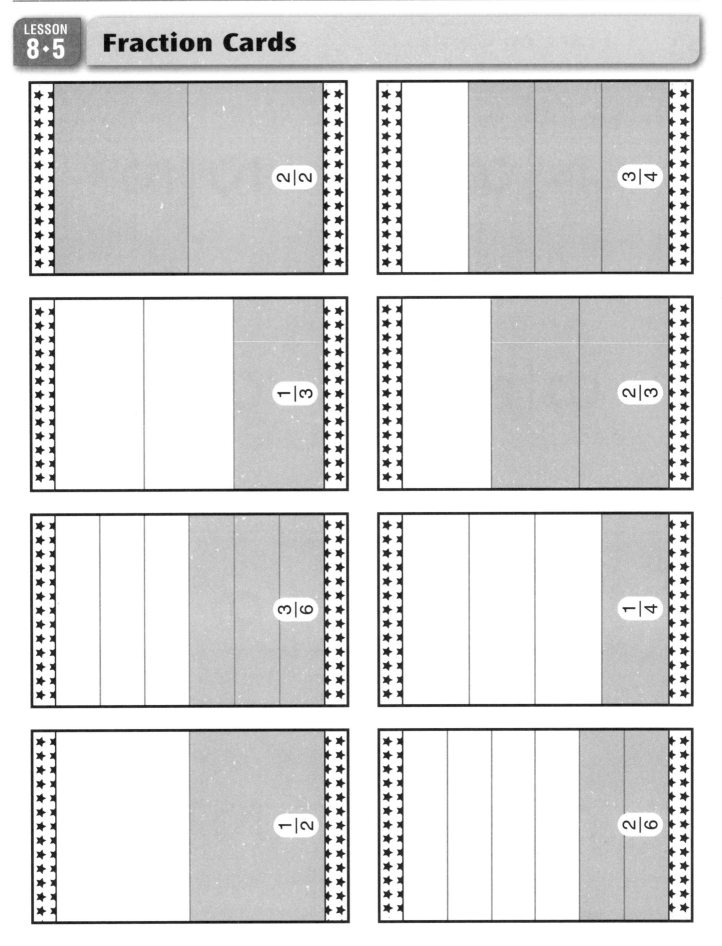

Date _____ Time _____

LESSON 8·5 **Fraction Cards**

$$\frac{3}{4}$$

$$\frac{2}{2}$$

$$\frac{2}{3}$$

$$\frac{1}{3}$$

$$\frac{1}{4}$$

$$\frac{3}{6}$$

$$\frac{2}{6}$$

$$\frac{1}{2}$$

Back of Activity Sheet 5

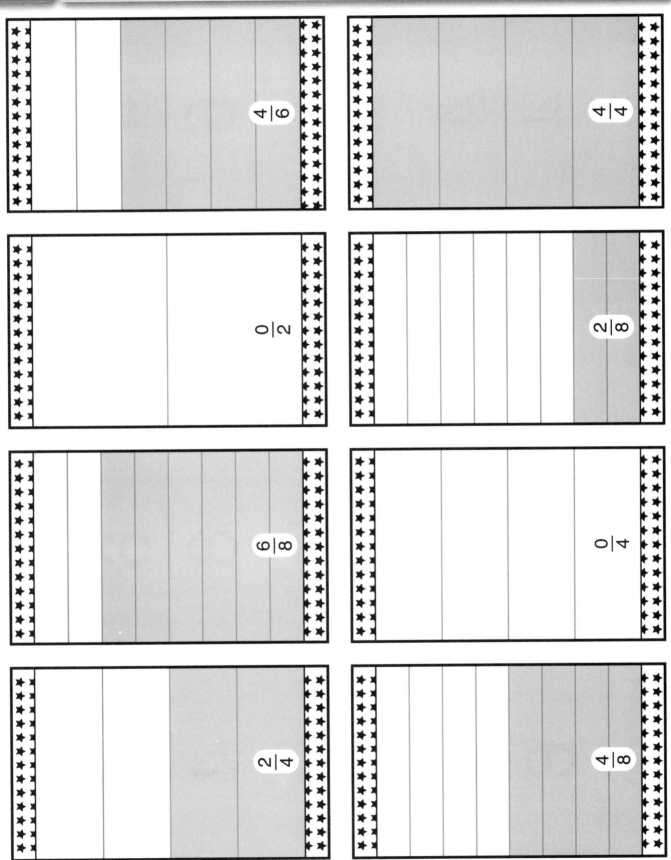

LESSON 8·5 Fraction Cards

(Cards shown rotated/upside-down:)

$\frac{4}{4}$ $\frac{4}{6}$

$\frac{2}{8}$ $\frac{0}{2}$

$\frac{0}{4}$ $\frac{6}{8}$

$\frac{4}{8}$ $\frac{2}{4}$

Back of Activity Sheet 6

Lesson 8·5 Fraction Cards

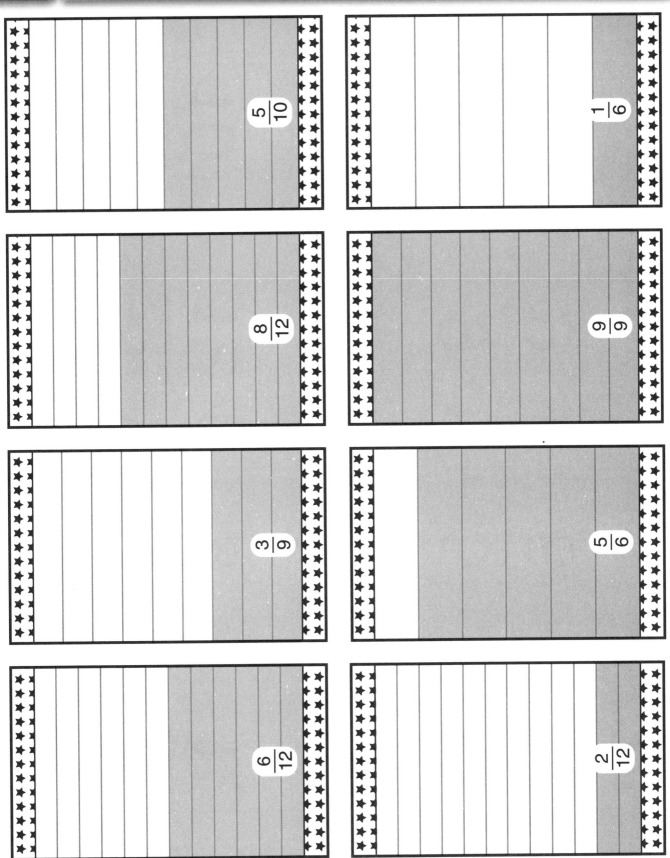

Activity Sheet 7

LESSON 8·5 Fraction Cards

$\dfrac{5}{10}$　　$\dfrac{1}{6}$

$\dfrac{8}{12}$　　$\dfrac{9}{9}$

$\dfrac{3}{9}$　　$\dfrac{5}{6}$

$\dfrac{6}{12}$　　$\dfrac{2}{12}$

Back of Activity Sheet 7

LESSON 8·5 Fraction Cards

$\frac{1}{5}$

$\frac{4}{12}$

$\frac{6}{9}$

$\frac{4}{5}$

$\frac{5}{5}$

$\frac{10}{12}$

$\frac{2}{10}$

$\frac{8}{10}$

Activity Sheet 8

LESSON 8·5 **Fraction Cards**

$\dfrac{1}{5}$

$\dfrac{4}{12}$

$\dfrac{6}{9}$

$\dfrac{4}{5}$

$\dfrac{5}{5}$

$\dfrac{10}{12}$

$\dfrac{2}{10}$

$\dfrac{8}{10}$

Back of Activity Sheet 8

TEACHER GUIDE

7th–9th Grade | Includes Student Worksheets | Science

- Weekly Lesson Schedule
- Worksheets
- Quizzes & Tests
- Answer Keys

Introduction to Anatomy & Physiology 2

First printing: September 2018

Copyright © 2018 by Master Books®. All rights reserved. No part of this book may be used or reproduced in any manner whatsoever without written permission of the publisher, except in the case of brief quotations in articles and reviews.
For information write:

Master Books®, P.O. Box 726, Green Forest, AR 72638

Master Books® is a division of the New Leaf Publishing Group, Inc.

ISBN: 978-1-68344-116-8
ISBN: 978-1-61458-681-4 (digital)

Unless otherwise noted, Scripture quotations are from the New King James Version of the Bible.

Printed in the United States of America

Please visit our website for other great titles:
www.masterbooks.com

For information regarding author interviews,
please contact the publicity department at (870) 438-5288.

Permission is granted for copies of reproducible pages from this text to be made for use within your own homeschooling family activities. Material may not be posted online, distributed digitally, or made available as a download. Permission for any other use of the material must be requested prior to use by email to the publisher at info@nlpg.com.

> Your reputation as a publisher is stellar. It is a blessing knowing anything I purchase from you is going to be worth every penny!
> —Cheri ★★★★★

> Last year we found Master Books and it has made a HUGE difference.
> —Melanie ★★★★★

> We love Master Books and the way it's set up for easy planning!
> —Melissa ★★★★★

> You have done a great job. MASTER BOOKS ROCKS!
> —Stephanie ★★★★★

> Physically high-quality, Biblically faithful, and well-written.
> —Danika ★★★★★

> Best books ever. Their illustrations are captivating and content amazing!
> —Kathy ★★★★★

Affordable
Flexible
Faith Building

Table of Contents

Using This Teacher Guide 4
Course Objectives 4
Course Description 5
Suggested Daily Schedule 7
Worksheets for *The Nervous System* 15
Worksheets for *The Digestive System & Metabolism* 55
Quizzes and Tests 95
Answer Keys 111

Dr. Tommy Mitchell has been a speaker and writer for Answers in Genesis since 2005. He has a degree in cell biology, as well as a medical degree. Once an evolutionist, now a creationist, he feels extremely passionate about sharing the vital creation/gospel message with the world, especially with influential teens.

Using This Teacher Guide

Features: The suggested weekly schedule enclosed has easy-to-manage lessons that guide the reading, worksheets, and all assessments. The pages of this guide are perforated and three-hole punched so materials are easy to tear out, hand out, grade, and store. Teachers are encouraged to adjust the schedule and materials needed in order to best work within their unique educational program.

Lesson Scheduling: Students are instructed to read the pages in their book and then complete the corresponding section provided by the teacher. Assessments that may include worksheets, activities, quizzes, and tests are given at regular intervals with space to record each grade. Space is provided on the weekly schedule for assignment dates, and flexibility in scheduling is encouraged. Teachers may adapt the scheduled days per each unique student situation. As the student completes each assignment, this can be marked with an "X" in the box.

🕐	Approximately 30 to 45 minutes per lesson, five days a week
🔑	Includes answer keys for worksheets, quizzes, and tests
📝	Worksheets for each section
🔄	Quizzes and tests are included to help reinforce learning and provide assessment opportunities
📄	Designed for grades 7 to 9 in a one-year course to earn 1 science credit

Course Objectives: Students completing this course will

- Learn how nerve signals are generated throughout the body
- Identify how nerve signals are transmitted to and from the brain
- Investigate the structure of the brain and how it processes input from the body
- Explore our senses: sight, hearing, taste, and more
- Discover the process of digestion by which the food we take in is converted to the substances our bodies need
- Learn about metabolism, the chemical transformations that happen in our cells

Course Description

The introduction to anatomy and physiology continues as students are given a deeper understanding of God's wonderful design of their bodies. How do just the correct muscles know how to contract in just the right way to allow us to walk? How can we control the movements of our hands in a very precise fashion so that we can brush our teeth? How can we decipher those funny marks on a printed page, understand that they are letters and punctuation marks, and make sense of them? How can we hear others singing and make our voices match theirs? How does the cereal you had for breakfast become energy? Or the popcorn you had at the ballgame? How does the chicken you had for supper provide the amino acids the body needs to build proteins? These questions and more are answered as we look into the wonders of God's awesome creation.

Our minds and bodies process vast amounts of information each second, information that comes from all parts of the body. Then nerve signals are sent out in response to those inputs. If this sounds simple, rest assured, it is not. It is all quite extraordinary! But as with all things in our fallen cursed world, things do go wrong. We will also explore the problems that occur when our bodies are damaged by disease or injury.

When you see the incredible complexity of you, you will realize that our bodies cannot be the result of chemical accidents occurring over millions of years. The human body is the greatest creation of an all-knowing Master Designer!

Note for Grading: All worksheet answers are worth 4 point each (100 points total) and quizzes and tests are 100 points total with all answers valued at 5 points each.

First Semester Suggested Daily Schedule

Date	Day	Assignment	Due Date	✓	Grade
		First Semester-First Quarter — *The Nervous System*			
Week 1	Day 1	Read Pages 4–6 (to Overview of the Nervous System) • *The Nervous System* (NS) • Read Introduction with focus on course objectives • Pages 4–5 • Teacher Guide (TG)			
	Day 2	Read Pages 6–9 (from Overview of Nervous System) • (NS)			
	Day 3	Worksheet 1 • Pages 17–18 • (TG)			
	Day 4	Worksheet 1 • Pages 17–18 • (TG)			
	Day 5	Read Structure of Nervous System • Pages 10–13 • (NS)			
Week 2	Day 6	Read Pages 14–16 • (NS)			
	Day 7	Worksheet 2 • Pages 19–20 • (TG)			
	Day 8	Worksheet 2 • Pages 19–20 • (TG)			
	Day 9	Read Pages 17–19 (to Nerves) • (NS)			
	Day 10	Read Pages 19–21 (from Nerves) • (NS)			
Week 3	Day 11	Worksheet 3 • Pages 21–22 • (TG)			
	Day 12	Worksheet 3 • Pages 21–22 • (TG)			
	Day 13	Read Pages 22–23 • (NS)			
	Day 14	Read Pages 24–25 (to The Action Potential) • (NS)			
	Day 15	Read Pages 25–27 (from The Action Potential) • (NS)			
Week 4	Day 16	Read Pages 28–29 • (NS)			
	Day 17	Read Pages 30–31 • (NS)			
	Day 18	Read Pages 32–34 (to The Role of the Synapse) • (NS)			
	Day 19	Read Pages 34–35 (from The Role of the Synapse) • (NS)			
	Day 20	Worksheet 4 • Pages 23–24 • (TG)			
Week 5	Day 21	Worksheet 4 • Pages 23–24 • (TG)			
	Day 22	Worksheet 4 • Pages 23–24 • (TG)			
	Day 23	Study Day			
	Day 24	**Quiz One** • Pages 97–98 • (TG)			
	Day 25	Read Pages 36–39 (to Cerebrospinal Fluid) • (NS)			
Week 6	Day 26	Read Pages 39–42 (from Cerebrospinal Fluid to Cerebrum - Gross Anatomy) • (NS)			
	Day 27	Worksheet 5 • Pages 25–26 • (TG)			
	Day 28	Worksheet 5 • Pages 25–26 • (TG)			
	Day 29	Read Pages 42–44 (from Cerebrum - Gross Anatomy to end of first paragraph) • (NS)			
	Day 30	Read Pages 44–46 (from first full paragraph to The Cerebrum) • (NS)			

Date	Day	Assignment	Due Date	✓	Grade
Week 7	Day 31	Worksheet 6 • Pages 27–28 • (TG)			
	Day 32	Worksheet 6 • Pages 27–28 • (TG)			
	Day 33	Read Pages 46–47 (from The Cerebrum to Cerebrum - Association Areas) • (NS)			
	Day 34	Read Pages 47–49 (from Cerebrum - Association Areas to Which Is the Important Side?) • (NS)			
	Day 35	Worksheet 7 • Pages 29–30 • (TG)			
Week 8	Day 36	Worksheet 7 • Pages 29–30 • (TG)			
	Day 37	Read Pages 49–51 (from Which Is the Important Side? to Brain Stem) • (NS)			
	Day 38	Read Pages 51–53 (from Brain Stem to Cerebellum) • (NS)			
	Day 39	Worksheet 8 • Pages 31–32 • (TG)			
	Day 40	Worksheet 8 • Pages 31–32 • (TG)			
Week 9	Day 41	Read Pages 53–55 (from Cerebellum to end of first paragraph) • (NS)			
	Day 42	Read Pages 55–56 (from second paragraph to Blood Brain Barrier) • (NS)			
	Day 43	Worksheet 9 • Pages 33–34 • (TG)			
	Day 44	Worksheet 9 • Pages 33–34 • (TG)			
	Day 45	Read Pages 56–57 (from Blood Brain Barrier) • (NS)			
		First Semester-Second Quarter			
Week 1	Day 46	Read Pages 58–59 (to Consciousness and the Mind) • (NS)			
	Day 47	Worksheet 10 • Pages 35–36 • (TG)			
	Day 48	Worksheet 10 • Pages 35–36 • (TG)			
	Day 49	Read Pages 59–61 (from Consciousness and the Mind to Spinal Cord - Gross Anatomy) • (NS)			
	Day 50	Read Pages 61–63 (from Spinal Cord - Gross Anatomy) • (NS)			
Week 2	Day 51	Worksheet 11 • Pages 37–38 • (TG)			
	Day 52	Worksheet 11 • Pages 37–38 • (TG)			
	Day 53	Read Pages 64–65 (to Tracts in the Spinal Cord) • (NS)			
	Day 54	Read Page 65 (from Tracts in the Spinal Cord) • (NS)			
	Day 55	Worksheet 12 • Pages 39–40 • (TG)			
Week 3	Day 56	Worksheet 12 • Pages 39–40 • (TG)			
	Day 57	Read Pages 66–68 • (NS)			
	Day 58	Read Pages 69–71 • (NS)			
	Day 59	Worksheet 13 • Pages 41–42 • (TG)			
	Day 60	Worksheet 13 • Pages 41–42 • (TG)			
Week 4	Day 61	Read Pages 72–73 • (NS)			
	Day 62	Read Pages 74–76 (to Sensory Receptors) • (NS)			
	Day 63	Read Pages 76–77 (from Sensory Receptors) • (NS)			
	Day 64	Read Pages 78–81 • (NS)			
	Day 65	Worksheet 14 • Pages 43–44 • (TG)			

Date	Day	Assignment	Due Date	✓	Grade
Week 5	Day 66	Worksheet 14 • Pages 43–44 • (TG)			
	Day 67	Study Day			
	Day 68	**Quiz Two** • Pages 99–100 • (TG)			
	Day 69	Read Pages 82–83 • (NS)			
	Day 70	Read Pages 84–85 • (NS)			
Week 6	Day 71	Worksheet 15 • Page 45 • (TG)			
	Day 72	Read Page 86 • (NS)			
	Day 73	Read Pages 87–88 (to Hearing) • (NS)			
	Day 74	Worksheet 16 • Pages 47–48 • (TG)			
	Day 75	Worksheet 16 • Pages 47–48 • (TG)			
Week 7	Day 76	Read Pages 88–90 (from Hearing to end of first paragraph) • (NS)			
	Day 77	Read Pages 90–91 (from start of second paragraph to Sound) • (NS)			
	Day 78	Worksheet 17 • Pages 49–50 • (TG)			
	Day 79	Worksheet 17 • Pages 49–50 • (TG)			
	Day 80	Read Pages 91–92 (from Sound) • (NS)			
Week 8	Day 81	Read Pages 93–95 • (NS)			
	Day 82	Worksheet 18 • Pages 51–52 • (TG)			
	Day 83	Worksheet 18 • Pages 51–52 • (TG)			
	Day 84	Read Pages 96–98 (to The Retina) • (NS)			
	Day 85	Read Pages 98–102 (from The Retina) • (NS)			
Week 9	Day 86	Worksheet 19 • Page 53 • (TG)			
	Day 87	Study Day			
	Day 88	**Quiz Three** • Pages 101–102 • (TG)			
	Day 89	Study Day			
	Day 90	**Test One** • Pages 107–108 • (TG)			
		Mid-Term Grade			

Second Semester Suggested Daily Schedule

Date	Day	Assignment	Due Date	✓	Grade
		Second Semester-Third Quarter — *The Digestion System & Metabolism*			
Week 1	Day 91	Read Foundations • Pages 4–7 • *Digestive System & Metabolism* (DSM)			
	Day 92	Read Pages 8–10 • (DSM)			
	Day 93	Read Pages 11–13 • (DSM)			
	Day 94	Worksheet 20 • Pages 57–58 • (TG)			
	Day 95	Worksheet 20 • Pages 57–58 • (TG)			
Week 2	Day 96	Read Pages 13–16 (to The Tongue) • (DSM)			
	Day 97	Read Pages 16–18 (from The Tongue) • (DSM)			
	Day 98	Worksheet 21 • Pages 59–60 • (TG)			
	Day 99	Worksheet 21 • Pages 59–60 • (TG)			
	Day 100	Read Pages 19–20 (to final full paragraph) • (DSM)			
Week 3	Day 101	Read Pages 20–23 (from final full paragraph) • (DSM)			
	Day 102	Worksheet 22 • Pages 61–62 • (TG)			
	Day 103	Worksheet 22 • Pages 61–62 • (TG)			
	Day 104	Read Pages 24–25 (to Saliva) • (DSM)			
	Day 105	Read Pages 25–27 (from Saliva) • (DSM)			
Week 4	Day 106	Worksheet 23 • Pages 63–64 • (TG)			
	Day 107	Worksheet 23 • Pages 63–64 • (TG)			
	Day 108	Read Pages 28–31 • (DSM)			
	Day 109	Read Pages 32–34 (to The Stomach) • (DSM)			
	Day 110	Worksheet 24 • Pages 65–66 • (TG)			
Week 5	Day 111	Worksheet 24 • Pages 65–66 • (TG)			
	Day 112	Read Pages 34–37 (from The Stomach) • (DSM)			
	Day 113	Read Pages 38–39 • (DSM)			
	Day 114	Worksheet 25 • Pages 67–68 • (TG)			
	Day 115	Worksheet 25 • Pages 67–68 • (TG)			
Week 6	Day 116	Read Pages 40–43 (to And Now...) • (DSM)			
	Day 117	Read Pages 43–44 (from And Now...) • (DSM)			
	Day 118	Worksheet 26 • Pages 69–70 • (TG)			
	Day 119	Worksheet 26 • Pages 69–70 • (TG)			
	Day 120	Read Pages 45–46 • (DSM)			
Week 7	Day 121	Read Pages 47–48 • (DSM)			
	Day 122	Worksheet 27 • Pages 71–72 • (TG)			
	Day 123	Worksheet 27 • Pages 71–72 • (TG)			
	Day 124	Read Pages 49–51 (to Blood Supply of the Liver) • (DSM)			
	Day 125	Read Pages 51–54 (from Blood Supply of the Liver to Functions of the Liver) • (DSM)			

Date	Day	Assignment	Due Date	✓	Grade
Week 8	Day 126	Worksheet 28 • Pages 73–74 • (TG)			
	Day 127	Worksheet 28 • Pages 73–74 • (TG)			
	Day 128	Read Pages 54–55 (from Functions of the Liver) • (DSM)			
	Day 129	Read Pages 56–57 • (DSM)			
	Day 130	Read Pages 58–59 • (DSM)			
Week 9	Day 131	Read Pages 60–61 (to Blood Supply of the Small Intestine) • (DSM)			
	Day 132	Worksheet 29 • Pages 75–76 • (TG)			
	Day 133	Worksheet 29 • Pages 75–76 • (TG)			
	Day 134	Study Day			
	Day 135	**Quiz One** • Pages 103–104 • (TG)			
		Second Semester-Fourth Quarter			
Week 1	Day 136	Read Pages 61–63 (from Blood Supply of the Small Intestine) • (DSM)			
	Day 137	Read Pages 64–65 • (DSM)			
	Day 138	Worksheet 30 • Pages 77–78 • (TG)			
	Day 139	Worksheet 30 • Pages 77–78 • (TG)			
	Day 140	Read Pages 66–67 • (DSM)			
Week 2	Day 141	Read Pages 68–69 • (DSM)			
	Day 142	Worksheet 31 • Pages 79–80 • (TG)			
	Day 143	Worksheet 31 • Pages 79–80 • (TG)			
	Day 144	Read Pages 70–73 • (DSM)			
	Day 145	Read Page 74 • (DSM)			
Week 3	Day 146	Worksheet 32 • Pages 81–82 • (TG)			
	Day 147	Worksheet 32 • Pages 81–82 • (TG)			
	Day 148	Read Pages 75–76 • (DSM)			
	Day 149	Read Pages 77–79 (to Digestion of Proteins) • (DSM)			
	Day 150	Worksheet 33 • Pages 83–84 • (TG)			
Week 4	Day 151	Worksheet 33 • Pages 83–84 • (TG)			
	Day 152	Read Pages 79–81 (from Digestion of Proteins to Lipids) • (DSM)			
	Day 153	Read Pages 81–83 (from Lipids) • (DSM)			
	Day 154	Worksheet 34 • Pages 85–86 • (TG)			
	Day 155	Worksheet 34 • Pages 85–86 • (TG)			
Week 5	Day 156	Read Pages 84–85 • (DSM)			
	Day 157	Read Pages 86–87 • (DSM)			
	Day 158	Worksheet 35 • Pages 87–88 • (TG)			
	Day 159	Worksheet 35 • Pages 87–88 • (TG)			
	Day 160	Read Pages 88–89 (to Water) • (DSM)			

Date	Day	Assignment	Due Date	✓	Grade
Week 6	Day 161	Read Pages 89–91 (from Water to Fiber) • (DSM)			
	Day 162	Worksheet 36 • Pages 89–90 • (TG)			
	Day 163	Worksheet 36 • Pages 89–90 • (TG)			
	Day 164	Worksheet 36 • Pages 89–90 • (TG)			
	Day 165	Read Pages 91–94 (from Fiber) • (DSM)			
Week 7	Day 166	Read Pages 95–96 • (DSM)			
	Day 167	Read Pages 97–98 • (DSM)			
	Day 168	Worksheet 37 • Pages 91–92 • (TG)			
	Day 169	Worksheet 37 • Pages 91–92 • (TG)			
	Day 170	Worksheet 37 • Pages 91–92 • (TG)			
Week 8	Day 171	Read Pages 99–102 (to Lipid Metabolism) • (DSM)			
	Day 172	Read Pages 102–105 (from Lipid Metabolism) • (DSM)			
	Day 173	Worksheet 38 • Pages 93–94 • (TG)			
	Day 174	Worksheet 38 • Pages 93–94 • (TG)			
	Day 175	Worksheet 38 • Pages 93–94 • (TG)			
Week 9	Day 176	Study Day			
	Day 177	**Quiz Two** • Pages 105–106 • (TG)			
	Day 178	Study Day Volume One			
	Day 179	Study Day Volume Two			
	Day 180	**Test Two** • Pages 109–110 • (TG)			
		Final Grade			

Nervous System Worksheets
for Use with
The Nervous System

| The Nervous System | Pages 4–9 | Day 3–4 | Worksheet 1 | Name |

Words to Know — Define the Following:

1. Sensory function:

2. Motor output:

3. Central nervous system:

4. The brain:

5. Peripheral nervous system:

6. Sensory division:

7. Afferent division:

8. Motor division:

9. Somatic nervous system:

10. Autonomic nervous system:

Fill in the Blank

1. The basic pattern of the nervous system consists of information coming into the nervous system. This information is then recognized and _____, and finally a signal is sent out instructing an organ (or organs) to respond in some manner.

2. The nervous system often compares what is sensed in the present to what has been _____ in the past.

3. _____ implies movement or some sort of action.

4. The two major divisions of the nervous system are the central nervous system (CNS) and the _____ nervous system (PNS).

5. The _____ cord extends from the base of the brain down to the lower levels of the spinal column.

6. The peripheral nervous system consists of the _____ nerves that extend from the brain, and the spinal nerves that extend from the spinal cord.

7. The PNS has two basic functions: carrying sensory information to the CNS and transmitting _____ out to the various part of the body.

Introduction to Anatomy & Physiology 2

8. We can divide the PNS into two divisions, which are the _____ division and the motor division.

9. The motor division is sometimes called the _____ (meaning "carrying away") division because it carries instructions "away from" the CNS.

10. _____ means "body," so this part of the nervous system allows us to control our body's movements.

Complete the Chart — Automatic Nervous System

Parasympathetic

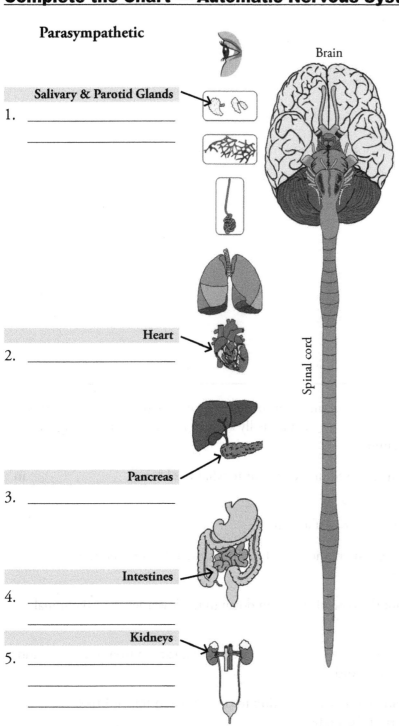

Salivary & Parotid Glands
1. _____
   _____

Heart
2. _____

Pancreas
3. _____

Intestines
4. _____
   _____

Kidneys
5. _____
   _____
   _____

| The Nervous System | Pages 10–16 | Day 7–8 | Worksheet 2 | Name

Words to Know — Define the Following:

1. Neurons:

2. Stimulus:

3. Neuroglia:

4. Neurotransmitters:

5. Dendrites:

6. Axon terminals:

7. Multipolar neurons:

8. Bipolar neurons:

9. Unipolar neurons:

10. Interneurons:

Fill in the Blank

1. _____ tissue lines body cavities or covers surfaces.

2. _____ tissue helps provide a framework for the body and helps connect and support other organs in the body.

3. _____ tissue is the primary component of the nervous system.

4. _____ tissue is responsible for movement and includes skeletal, smooth, and cardiac.

5. The _____ is composed of three parts: the cell body, dendrites, and the axon.

6. The _____ is the portion of the neuron that carries a nerve impulse away from the cell body.

7. Unlike most cell types in your body, neurons cannot be routinely _____.

Introduction to Anatomy & Physiology 2

8. Neurons that transmit impulses away from the central nervous system are called motor or _____ neurons.

9. Sensory or _____ neurons carry impulses triggered by sensory receptors toward the central nervous system.

10. The four types of _____ cells in the central nervous system are the astrocytes, microglial cells, ependymal cells, and oligodendrocytes.

Complete the Chart — Neuron

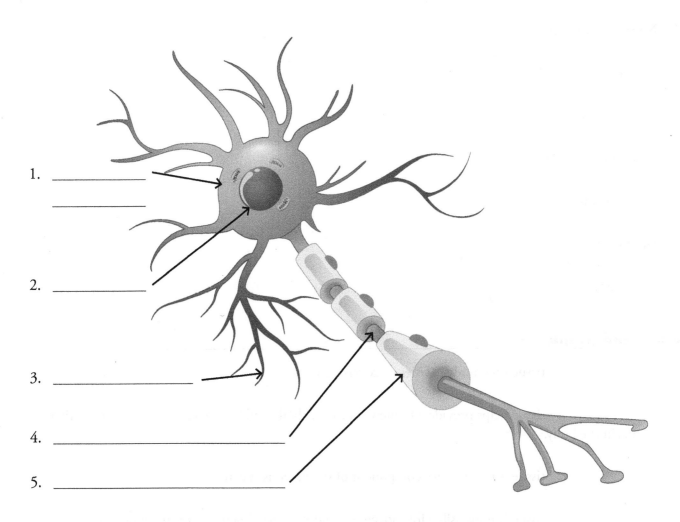

1. _____
2. _____
3. _____
4. _____
5. _____

The Nervous System | Pages 17–21 | Day 11–12 | Worksheet 3 | Name

Words to Know — Define the Following:

1. Myelination:

2. Schwann cells:

3. Multiple Sclerosis (MS):

4. Neuron:

5. Motor neurons:

6. Sensory neurons:

7. Mixed nerves:

8. Nerve damage in the PNS:

9. Nerve damage in the CNS:

10. Wallerian degeneration:

Fill in the Blank

1. The _____ sheath provides electrical insulation for the axon.

2. There are small gaps between adjacent Schwann cells called _____ of Ranvier.

3. In the CNS, it is the oligodendrocyte that is responsible for _____.

4. The number of myelinated axons _____ from birth throughout childhood until adulthood.

5. Symptoms of MS include double vision, weakness, loss of _____, and paralysis.

6. A newborn baby has very little _____ of its body in the beginning.

7. A _____ is made of bundles of axons located in the peripheral nervous system.

8. With rare exceptions, mature neurons do not divide to _____ themselves.

9. We are not the products of chance, but special _____.

10. The enormous _____ of the body should remind us constantly of God's wisdom and creativity.

Complete the Chart — Anatomy of a Nerve

1. _____

2. _____

3. _____

4. _____

5. _____

The Nervous System — Worksheet 4

Pages 22–35 | Day 20–22

Words to Know — Define the Following:

1. Action potential:

2. Depolarization:

3. Threshold:

4. All-or-none event:

5. Repolarization:

6. Continuous conduction:

7. Saltatory conduction:

8. Graded potentials:

9. Synapse:

10. Chemical synapse:

Fill in the Blank

1. Your body itself is electrically _____, which means that the number of negative particles equals the number of positive particles.

2. K+ from the inside of the cell rushes out of the cell toward the area of lower concentration, which is called a _____ gradient.

3. There exists a small electrical difference across the cell membrane, which is called the _____ membrane potential.

4. Any depolarization not reaching threshold is sub-threshold and will not trigger a nerve _____.

5. When the charge across the membrane lessens it is called _____.

6. A stimulus from sensory receptors sensing a strong wind will trigger more _____ action

potentials by comparison to those triggered by a mild breeze.

7. If the neurotransmitter causes the postsynaptic neuron to become more negative inside, this is called hyper-_____.

8. Electrical _____ directly transmit signals to adjacent cells.

9. Neurotransmitters are the molecules that carry the _____ across the synaptic cleft.

10. The body is constantly trying to maintain a "balance" among its many systems, which has come to be known as _____.

Complete the Chart — Synapses Can Occur in Many Locations

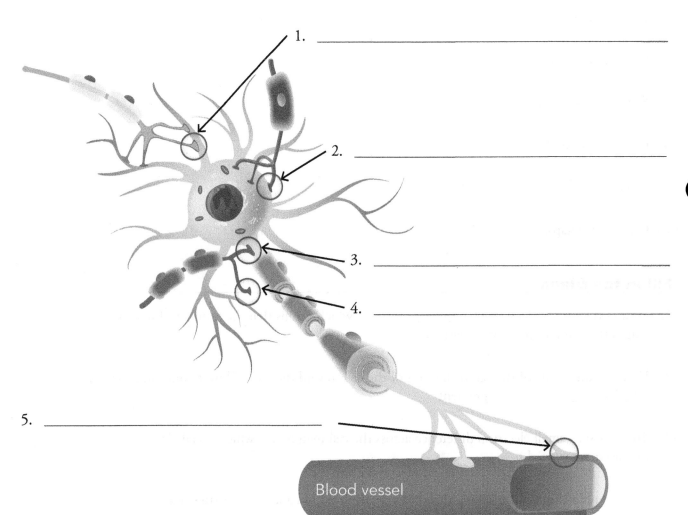

1. _____
2. _____
3. _____
4. _____
5. _____

The Nervous System — Worksheet 5

Pages 36–42 · Day 27–28

Words to Know — Define the Following:

1. Cranial vault:

2. Meninges:

3. Periosteal:

4. Meningeal:

5. Pia mater:

6. Cerebrospinal fluid (CSF):

7. Meningitis:

8. Cerebrum:

9. Sulci:

10. Cerebral hemispheres:

Fill in the Blank

1. An average brain weighs around _____ pounds.

2. Even though its size is small, the brain consumes about _____ percent of the body's total energy.

3. The number of _____ in the brain, links between its neurons, has been estimated to be in the trillions.

4. The _____ is the thickest layer of the meninges and is itself composed of two layers.

5. The _____ mater covers the brain and spinal cord.

6. The CSF supports the brain by helping it _____ in the cranial vault.

7. The interconnections between the four ventricles and the central canal allow the CSF to _____ around the brain and spinal cord.

8. The majority of the CSF is produced by the frond-like choroid plexus found in each _____.

9. The cerebrum is somewhat wrinkled in appearance, covered with many ridges and _____.

10. Several smaller fissures divide each hemisphere into smaller portions, called _____.

Complete the Chart — Ventricles of the Brain

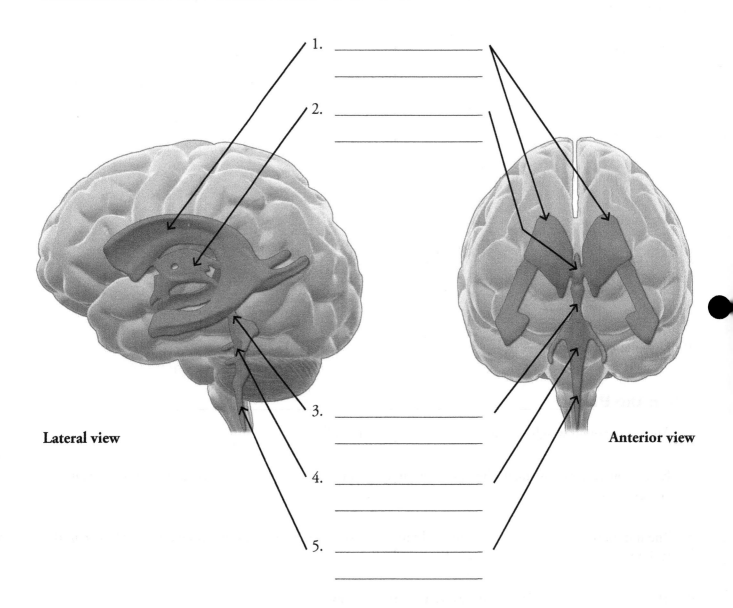

1. _____
   _____

2. _____
   _____

3. _____
   _____

4. _____
   _____

5. _____
   _____

Lateral view   **Anterior view**

The Nervous System — Pages 42–46 — Day 31–32 — Worksheet 6 — Name

Words to Know — Define the Following:

1. Gray matter:

2. White matter:

3. Corpus callosum:

4. Voluntary movement:

5. Lateralized (lateralization):

6. Precentral gyrus:

7. Pyramidal neurons:

8. Tracts:

9. Motor homunculus:

10. Premotor cortex:

Fill in the Blank

1. Because of the _____ that cross from one side of the cerebrum to the other through the corpus callosum, the two cerebral hemispheres communicate and work together.

2. We can remember the things we learn and experience because the cerebrum helps store _____.

3. The cerebral hemispheres control the _____ sides of the body.

4. The right _____ controls and monitors the left side of the body.

5. The signals into and out of the brain are constantly enhanced, inhibited, and modulated (_____) by input from all across the cerebrum.

6. Our _____ function is controlled by a region of the cerebrum called the primary motor cortex.

7. The hand and _____ take up the majority of the motor cortex.

8. The _____ muscle movements are fine-tuned by a vast number of neural inputs from areas throughout the cerebral cortex.

9. Broca's area is _____, so it is found in only one of the hemispheres.

10. Broca's area controls muscles involved with _____.

Complete the Chart — Sagittal Section of the Brain

1. _____

2. _____

3. _____

4. _____

5. _____

| The Nervous System | Pages 46–49 | Day 35–36 | Worksheet 7 | Name |

Words to Know — Define the Following:

1. Primary somatosensory cortex:

2. Expressive aphasia:

3. Somatosensory association cortex:

4. Association areas:

5. Frontal association area:

6. Visual association area:

7. Wernicke's area:

8. Auditory association area:

Fill in the Blank

1. Sensory information is _____ in the primary somatosensory cortex.

2. The primary somatosensory cortex is located in the _____ lobe of the cerebrum on the postcentral gyrus.

3. Touching a piano key, sitting in a chair, drinking warm cocoa, stubbing your toe . . . all these actions produce sensory inputs that are ultimately processed in the primary _____ cortex.

4. When people suffer damage to Broca's area, perhaps after a stroke, they often lose the ability to _____.

5. Our brain can _____ whether a round object in your hand is a marble or a grape, even without looking.

6. It said that the _____ hemisphere is the seat of logic, reasoning, and planning.

7. No one is truly that "_____" on either side of the brain.

Complete the Chart — Association Areas

1. _____

2. _____

3. _____

4. _____

5. _____

6. _____

7. _____

8. _____

9. _____

10. _____

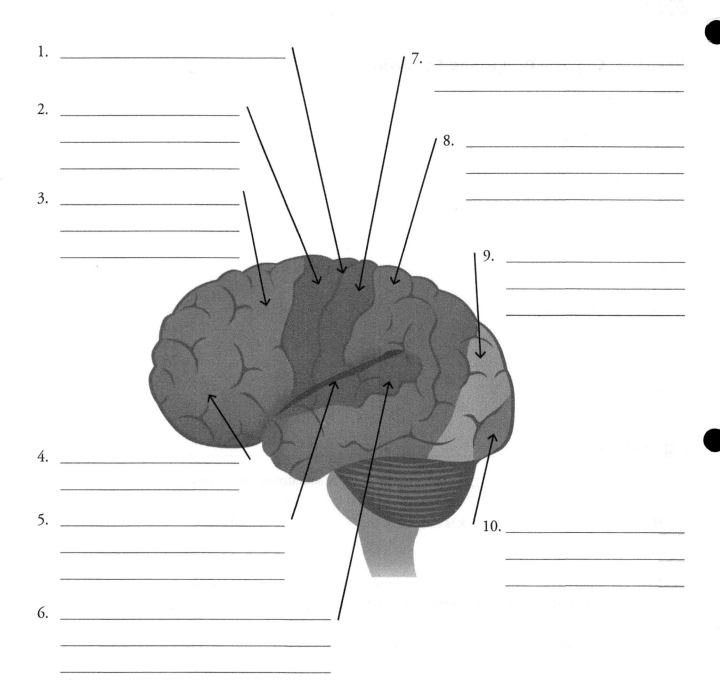

The Nervous System — Worksheet 8

Pages 49–53 | Day 39–40

Words to Know — Define the Following:

1. Diencephalon:

2. Thalamus:

3. Hypothalamus:

4. Homeostasis:

5. Brain stem:

6. Medulla oblongata:

7. Reticular formation:

Fill in the Blank

1. The white matter bands in the corpus callosum connect the right and left hemispheres to allow _____ between the hemispheres.

2. Your attentiveness is affected by the hypothalamus, as are your _____ cycles.

3. The _____ is located between the diencephalon and the pons.

4. The _____ are involved in both hearing and vision.

5. The pons (meaning "_____") is positioned between the midbrain and the medulla oblongata.

6. Just before reaching the spinal cord, most fibers in the pyramidal tracts _____ (cross over) to the opposite side.

7. The most commonly cited cause of tension headaches is _____.

8. _____ headaches are generally (but not always) more severe than tension headaches.

Complete the Chart — Hypothalamus and Its Function

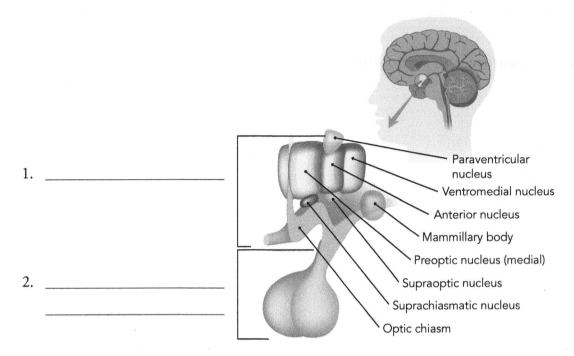

1. _____

2. _____

HYPOTHALAMIC NUCLEI	FUNCTION
Paraventricular nucleus	3.
Suprachiasmatic nucleus	4.
Ventromedial nucleus	5.
Anterior hypothalamic nucleus	6.
Mammillary body	7.
Supraoptic nucleus	8.
Posterior nucleus	9.
Tuberomammillary nucleus	10.

32 // Introduction to Anatomy & Physiology 2

The Nervous System | Pages 53–56 | Day 43–44 | Worksheet 9 | Name

Words to Know — Define the Following:

1. Cerebellum:

2. Subclavian arteries:

3. Subclavian steal syndrome:

4. Circle of Willis:

5. Stroke:

6. Ischemic stroke:

7. Aneurysm:

Fill in the Blank

1. The cerebellum and brainstem are connected by three bundles of white matter called _____.

2. The cerebellum is responsible for making you aware of your _____.

3. Even though the brain makes up only about 2 percent of the body's weight, it requires about _____ percent of the body's oxygen and glucose.

4. A loss of oxygen for as little as _____ minutes can result in permanent brain damage.

5. If you press on your neck gently on either side of your Adam's apple, the pulse you feel is from your internal _____ artery.

6. To help ensure the brain's _____ blood supply, the right and left anterior cerebral arteries are connected by the anterior communicating artery.

7. Strokes can result in _____ neurologic damage or even death.

8. _____ can result from the buildup of atherosclerotic plaques in an artery or a blood clot that blocks blood flow.

9. A _____ stroke is caused by bleeding directly into the space around the brain.

10. Circulation to the _____ can be maintained by blood flow coming from the other arteries making up the circle of Willis.

Complete the Chart — Blood Supply to the Brain

1. _____

2. _____

3. _____

4. _____

5. _____

6. _____

7. _____

8. _____

| The Nervous System | Pages 56–59 | Day 47–48 | Worksheet 10 | Name |

Words to Know — Define the Following:

1. Electrocardiogram (ECG):

2. Electroencephalogram (EEG):

3. Delta waves:

4. Theta waves:

5. Alpha waves:

6. Beta waves:

7. Sleep:

8. Rapid eye movement (REM):

9. Amnesia:

10. Synaptic plasticity:

Fill in the Blank

1. To prevent potentially harmful things from coming into contact with brain tissue, there is a blood brain _____.

2. Oxygen, glucose, and carbon dioxide can _____ pass across the blood brain barrier.

3. An EEG records the _____ activity of neurons close to the surface of the brain, those located in the cerebral cortex.

4. The various patterns of electrical activity seen on an EEG are called brain _____.

5. 1 Hz (1 Hertz) means one peak each _____.

6. In the tragic circumstance where there is no detectable brain activity on the EEG, the patient is said to be "brain _____."

7. A person passes through stages of non-_____ eye movement sleep as he or she falls deeper and deeper into sleep.

Introduction to Anatomy & Physiology 2 // 35

8. Most _____ occurs during REM sleep.

9. A vital part of learning is _____, since anything you learn is useless to you if you cannot recall it at the proper time.

10. Most often we must be exposed to _____ multiple times before we learn it.

Complete the Chart — Brain Waves

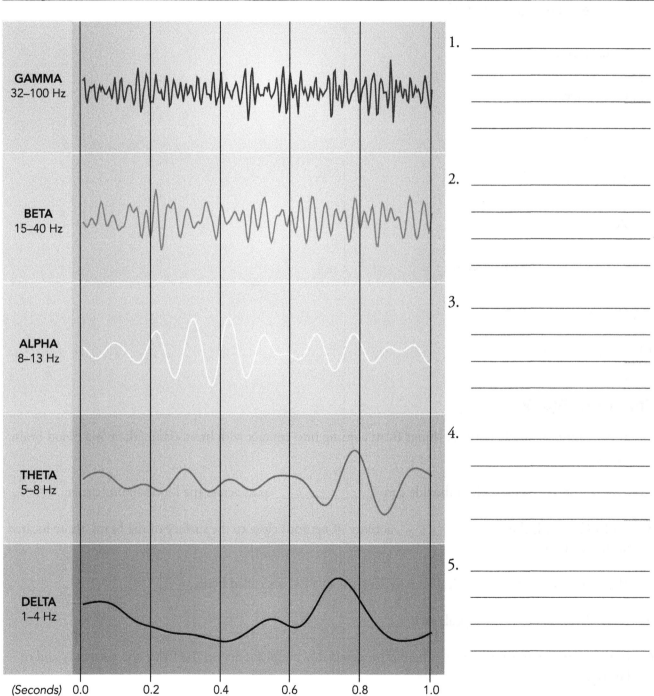

1. _____
2. _____
3. _____
4. _____
5. _____

| The Nervous System | Pages 59–63 | Day 51–52 | Worksheet 11 | Name |

Words to Know — Define the Following:

1. Consciousness:

2. Spinal cord:

3. Spinal nerves:

4. Intervertebral foramina (plural):

5. Window (foramen):

6. Spinal cord segment:

7. Horns:

8. Ganglion:

9. Interneurons:

10. Lateral horns:

Fill in the Blank

1. The human body was designed by the Creator God, and it is more than just a collection of _____.

2. Who has put _____ in the mind? Or who has given understanding to the heart? (Job 38:36)

3. Without the _____ cord the brain would be unable to receive sensory information from your body or tell your body what to do.

4. The spinal cord starts at the medulla oblongata and extends to roughly the level of the _____ lumbar vertebral bone in the lower back.

5. The naming of the spinal nerves is based on the level of the vertebral column near which they exit, not the level of the spinal cord at which they _____.

6. In the spinal cord, gray matter consists mostly of neuron cell bodies, and white matter consists mainly of _____.

7. The projections to the rear are the right and left dorsal horns (also called the _____ horns).

Introduction to Anatomy & Physiology 2 // 37

8. Signals from sensory neurons are _____ to interneurons.

9. Lateral horns contain cell bodies of _____ neurons in the autonomic nervous system.

10. The spinal cord's white matter is divided into _____.

Complete the Chart — Spinal Nerves

1. _____
2. _____
3. _____
4. _____
5. _____

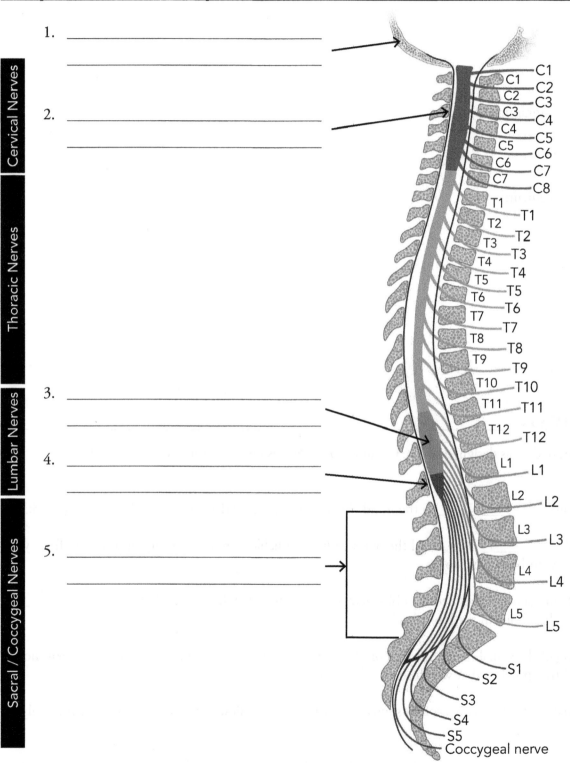

The Nervous System — Worksheet 12

Pages 64–65 | Day 55–56

Words to Know — Define the Following:

1. Ascending tracts:

2. Descending tracts:

3. Amyotrophic Lateral Sclerosis (ALS):

4. Sporadic ALS:

5. Familial ALS:

6. Roots:

7. Ventral root:

8. Dorsal root:

9. Corticospinal tract:

10. Spinothalamic tract:

Fill in the Blank

1. Sensory signals are sent to the spinal cord and then up (_____) to the brain.

2. At present there is no _____ for ALS.

3. Amyotrophic lateral sclerosis is also known as Lou _____ disease.

4. The dorsal root ganglion contains the cell bodies of sensory _____.

5. The ventral and dorsal roots _____ to form the 31 pairs of spinal nerves.

Complete the Chart — Spinal Tracts

1. _____
   _____

2. _____
   _____
   _____

   Fasciculus gracilis
   Fasciculus cuneatus

3. _____
   _____

   Dorsal
   Ventral

4. _____
   _____

   Lateral
   Ventral

5. _____
   _____

   Ventral white commissure

6. _____
   _____

   (pyramidal tracts)
   Lateral
   Ventral

7. _____
   _____

8. _____
   _____

   Medial
   Lateral

9. _____
   _____

10. _____
    _____

40 // Introduction to Anatomy & Physiology 2

| The Nervous System | Pages 66–71 | Day 59–60 | Worksheet 13 | Name |

Words to Know — Define the Following:

1. Peripheral nervous system:

2. Cranial nerves:

3. Olfactory (I) nerves:

4. Optic (II) nerves:

5. Trigeminal neuralgia:

6. Oculomotor nerve:

7. Bell's palsy:

8. Vagus nerve:

9. Hypoglossal nerve:

10. Plexuses:

Fill in the Blank

1. Your cranium has many little _____ through which your brain connects to the world using cranial nerves.

2. The _____ nerve is a sensory nerve.

3. The _____ nerve carries nerve impulses for vision.

4. Four of the _____ pairs of cranial nerves are devoted to your eyes.

5. Trigeminal neuralgia is characterized by episodes of _____ across the face in areas supplied by the trigeminal nerve.

6. Bell's palsy is caused by inflammation of the facial nerve, usually due to a _____ infection.

7. Each eye also has some intrinsic muscles — muscles located _____ the eye itself — that adjust the size of the pupil and focus the lens.

Introduction to Anatomy & Physiology 2 // 41

8. Spinal nerves carry sensory input to the CNS and motor _____ away from the CNS.

9. The four major plexuses are the cervical, brachial, _____, and sacral.

10. The major nerves that extend from the _____ plexus are the axillary nerve, the radial nerve, the median nerve, and the ulnar nerve.

Complete the Chart — The Brachial Plexus

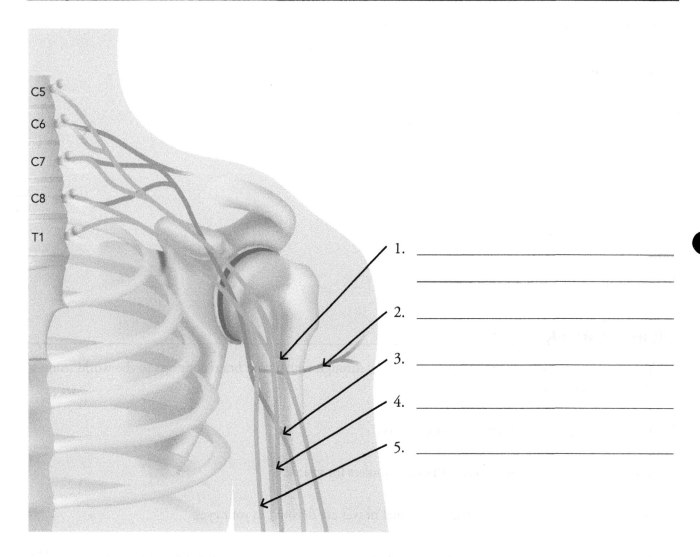

1. _____
2. _____
3. _____
4. _____
5. _____

The Nervous System — Worksheet 14

Pages 72–81 · Day 65–66

Words to Know — Define the Following:

1. The femoral nerve:

2. Carpal tunnel syndrome (CTS):

3. Shingles (herpes zoster):

4. Reflex:

5. Somatic reflex:

6. Autonomic reflex:

7. Autonomic nervous system (ANS):

8. Homeostasis:

9. Ganglion (plural, ganglia):

10. Parasympathetic nervous system:

Fill in the Blank

1. The _____ nerve supplies the muscles on the back of the thigh, the lower leg, and the foot.

2. As the _____ nerves control the movement of the diaphragm, injury to these nerves is a very severe situation.

3. The region of the body that provides sensory input to a particular spinal nerve (or segment) is called a _____.

4. With an _____ reflex, you remain unaware of what happened.

5. Sensory receptors in the nervous system allow us to be aware of changes in our environment, which are called _____.

6. Sensory receptors can be found in _____ muscle, skin, joints, and visceral organs.

7. There are two divisions to the autonomic nervous system, called _____ and parasympathetic.

8. Each motor pathway in the ANS is composed of two neurons, the _____ neuron and the postganglionic neuron.

9. Neurons of the somatic motor system form networks of fibers called _____, each located near the effector organ or region it serves.

10. The sympathetic nervous system is called the "_____ or flight" system.

Complete the Chart — The Fight or Flight Response

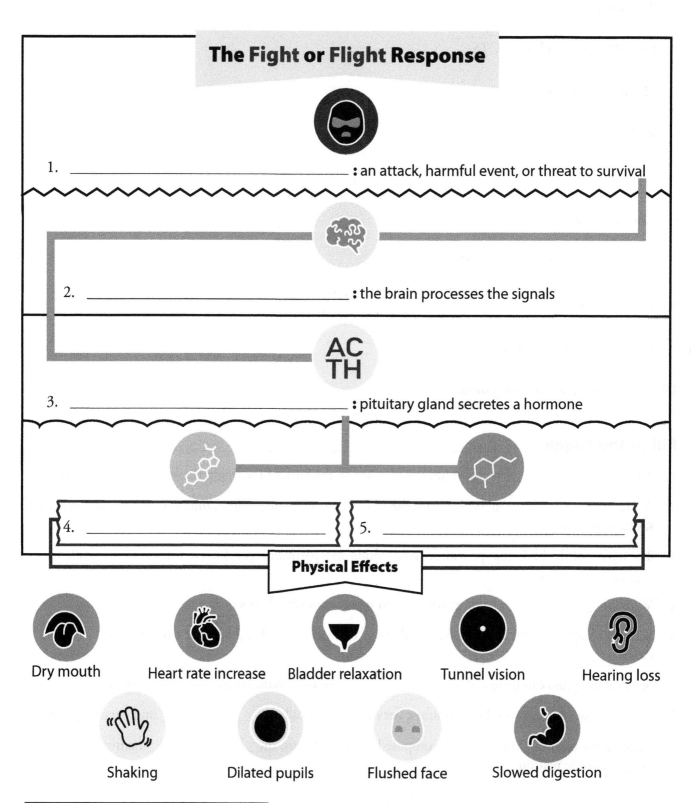

| The Nervous System | Pages 82–85 | Day 71 | Worksheet 15 | Name |

Words to Know — Define the Following:

1. Mechanoreceptors:

2. Chemoreceptors:

3. Photoreceptors:

4. Thermoreceptors:

5. Olfaction:

6. Cilia:

7. Basal cells (of an epithelium):

8. Odorants:

9. Triggering a sensory neuron:

10. Threshold for smell:

Fill in the Blank

1. The _____ senses involve sensory receptors contained in specialized organs or structures in the body, such as sight, hearing, and taste, which are examples of special senses.

2. Specialized _____ are contained in the retina of the eye.

3. The special senses are taste, sight, smell, hearing, and _____ (or equilibrium).

4. The _____ senses utilize sensory receptors scattered throughout the body.

5. Touch is certainly a _____ sense, but it is better to consider it more of a general sense.

6. The hearing ear and the seeing eye, The Lord has made them both. (_____ 20:12)

7. The dendrite of an olfactory neuron ends in several _____.

8. The bundles of _____ of the olfactory neurons extend through holes in the cribriform plate.

9. The basal cells of an epithelium are actually _____ cells that divide to produce replacement olfactory sensory neurons.

10. If enough odorant molecules trigger the _____ cell, an impulse is sent down the entire length of the neuron.

Complete the Chart — How Smelling Works

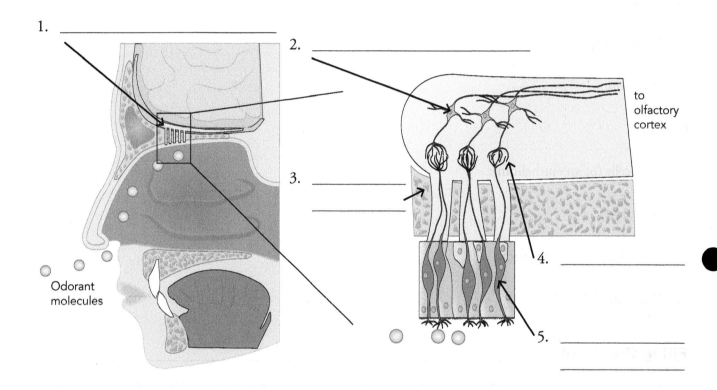

The Nervous System | Pages 86–88 | Day 74–75 | Worksheet 16 | Name

Words to Know — Define the Following:

1. Fungiform papillae:

2. Gustatory epithelial cells:

3. Tastant:

4. Afferent (sensory) fibers:

5. Bitter taste:

6. Sweet taste:

7. Sour taste:

8. Salty taste:

9. Umami:

10. Oleogustus:

Fill in the Blank

1. Our ability to taste depends on our ability to detect, and then react to, certain _____ in our environment.

2. Your tongue has a rather rough-looking surface due to the presence of many protuberances called _____ on its surface.

3. Taste _____ are found in the side walls of the foliate and circumvallate papillae.

4. In a taste bud, the _____ is not the sensory receptor.

5. The actual _____ cell for taste is the gustatory epithelial cell.

6. The gustatory epithelial cell releases a _____ that then stimulates receptors in the dendrites of the sensory neurons that are in the taste bud.

7. Every part of the _____ can detect any of the taste modalities.

8. If our _____ sense is triggered when we eat, it enhances the pleasure our food gives us.

Complete the Chart — Taste Buds

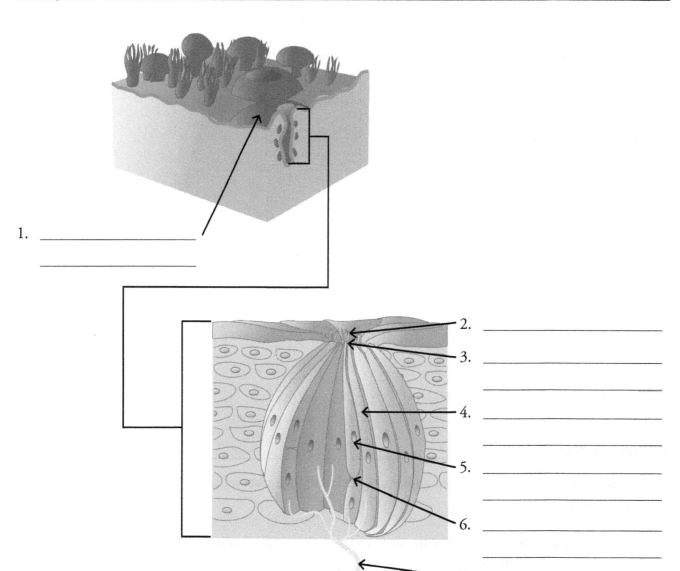

1. _____
   _____
2. _____
3. _____
4. _____
5. _____
6. _____
7. _____

The Nervous System — Worksheet 17

Words to Know — Define the Following:

1. Auricle:

2. Earwax:

3. Tympanic membrane:

4. The vestibule:

5. Cochlea:

Fill in the Blank

1. The ear is the organ associated with hearing and also the organ responsible for _____.

2. The ear is composed of three main regions: the external ear, the _____ ear, and the inner ear.

3. The external auditory canal is a tube through which sound waves move toward the tympanic membrane (_____).

4. The middle ear is a small cavity in the temporal _____ of the skull.

5. Strung across the middle ear is a chain of three small bones, which are the three smallest bones in the body: the malleus, the _____, and the stapes.

6. The _____ that start on the eardrum are going to be transmitted to the hammer, then to the anvil, then to the stirrup, and on to the oval window.

7. The inner ear has two parts, which are the bony labyrinth and the _____ labyrinth.

8. There are three major parts of the bony labyrinth, which are the vestibule, the semicircular _____, and the cochlea.

9. Inside the cochlear duct is the spiral organ, which is the hearing _____.

10. The cochlear duct is separated from the scala vestibuli by the _____ membrane.

Complete the Chart — Middle Ear

1. _____

2. _____
   _____

3. _____
   _____

4. _____
   _____

5. _____

6. _____

7. _____

8. _____

9. _____

10. _____
    _____

| The Nervous System | Pages 91–95 | Day 82–83 | Worksheet 18 | Name |

Words to Know — Define the Following:

1. Sound:

2. Pitch:

3. Hearing:

4. The external ear:

5. Basilar membrane:

6. Vestibular apparatus:

7. Utricle and saccule:

8. Macula:

9. Vertigo:

10. Labyrinthitis:

Fill in the Blank

1. When we hear something, we are sensing sound _____ from the environment.

2. The greater the _____, the louder the sound.

3. "I have heard of You by the hearing of the ear, But now my eye sees You." (_____ 42:5)

4. When it reaches the tympanic membrane, a sound wave strikes the membrane and causes it to vibrate at the same _____ as the sound wave.

5. The louder the sound reaching the tympanic membrane, the _____ the membrane is pushed as it vibrates.

6. When the tympanic membrane vibrates, it causes movement in the three _____ of the middle ear.

7. The vibrations transmitted from the _____ membrane are thus transferred to the oval window, and it begins to vibrate accordingly.

8. The basilar membrane is "_____" to respond to different frequencies along its length.

9. On the end nearer the oval window, the membrane responds better to _____ frequencies.

10. Other possible causes of vertigo include Ménière's disease, migraine, _____, multiple sclerosis, and Parkinsonism.

Complete the Chart — Auditory Path

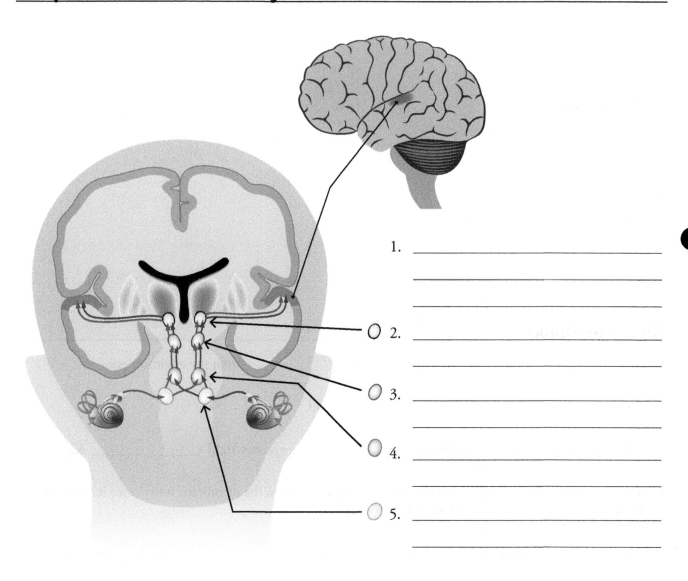

1. _____
2. _____
3. _____
4. _____
5. _____

| The Nervous System | Pages 96–102 | Day 86 | Worksheet 19 | Name |

Words to Know — Define the Following:

1. Corneal abrasion:

2. Glaucoma:

3. Iris:

4. Retina:

5. Cataract:

6. Müller cells:

7. Photoreceptors:

8. Color blindness:

9. Nearsighted:

10. Farsighted:

Fill in the Blank

1. The outside layer of the eye consists of two parts: the sclera and the _____.

2. The iris is a round, flat layer of smooth muscle with an opening in the middle called the _____.

3. _____ has to go all the way through the retina — past blood vessels and past two layers of neurons — to stimulate the photoreceptor cells so that a nerve signal can make its way back out of it.

4. The amount of _____ flowing in the choroid layer is very high, able to provide oxygen and nutrients needed by the photoreceptors.

5. The _____ are the receptors that are the most sensitive to light. Without them, we could not see in dim light or at night.

6. There are three types of _____: blue ones, which are obviously sensitive to blue light, red ones, which sense red light, and green ones, which detect green light.

Introduction to Anatomy & Physiology 2 // 53

7. In order for us to see things clearly, the light that enters the eye needs to be _____ sharply on the retina.

8. When light passes through a transparent object, it can be bent, or _____.

9. Some estimates state that as many as _____ percent of males have some degree of color blindness.

10. As people grow older, their lenses often become stiffer, and called _____.

Complete the Chart — Retina

1. _____

2. _____

3. _____

4. _____

5. _____

Digestive System & Metabolism Worksheets

for Use with

The Digestive System & Metabolism

Digestive System & Metabolism — Pages 4–13 — Day 94–95 — Worksheet 20 — Name

Words to Know — Define the Following:

1. Digestion:

2. Alimentary canal:

3. The accessory digestive organs:

4. Mechanical digestion:

5. Chemical digestion:

6. Absorption:

7. Elimination:

8. Serosa:

9. Peritoneum:

10. Bolus:

Fill in the Blank

1. The first function of the digestive system is called _____.

2. _____ is when food is moved along the length of the GI tract.

3. The indigestible material eliminated from the body is called _____ and leaves the body through the anus.

4. The food you chew up and swallow enters the _____, where it is processed and moved along from section to section.

5. The innermost tissue layer in the GI tract wall is called the _____.

6. The dense connective tissue of the _____ supports the overlying mucosa as it expands to accommodate food to be digested and shrinks back when digestion is completed.

Introduction to Anatomy & Physiology 2

7. _____ help secure organs to the body wall and hold them in the proper position so that they won't twist while also suspending them to allow them room to expand and to slide along other organs.

8. _____ is a condition resulting from an acute inflammation of the peritoneum.

9. Symptoms of peritonitis include _____ pain and fever.

10. The GI tract has its own nervous system, called the _____ nervous system.

Complete the Chart — Tissue Layers of the GI Tract

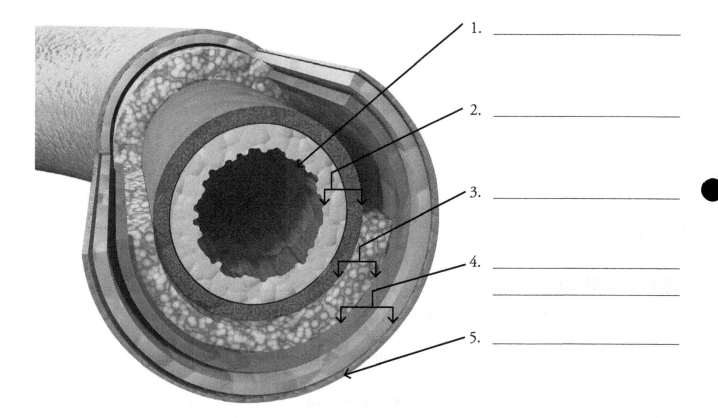

1. _____
2. _____
3. _____
4. _____
   _____
5. _____

Words to Know — Define the Following:

1. Hard palate:

2. Soft palate:

3. Papillae:

4. Tooth's neck:

5. Gingiva:

Fill in the Blank

1. The _____ are covered by skin on the outside but by mucous membrane on the inside of the mouth.

2. The lips, containing _____ muscle, are under voluntary control.

3. The superior (upper) boundary of the mouth is formed by the hard and soft palates, which is called the "_____" of the mouth.

4. The _____ is composed of two sets of skeletal muscles.

5. The tongue's extrinsic muscles are attached to the _____ bone.

6. _____ buds are found in fungiform, foliate, and circumvallate papillae.

7. The more thoroughly food is chewed, the better for your _____.

8. Each tooth has three major regions: the crown, the neck, and the _____.

9. _____ is the hardest substance in the body, and it is very durable.

10. _____ makes up the majority of the volume of a tooth.

Complete the Chart — The Tongue

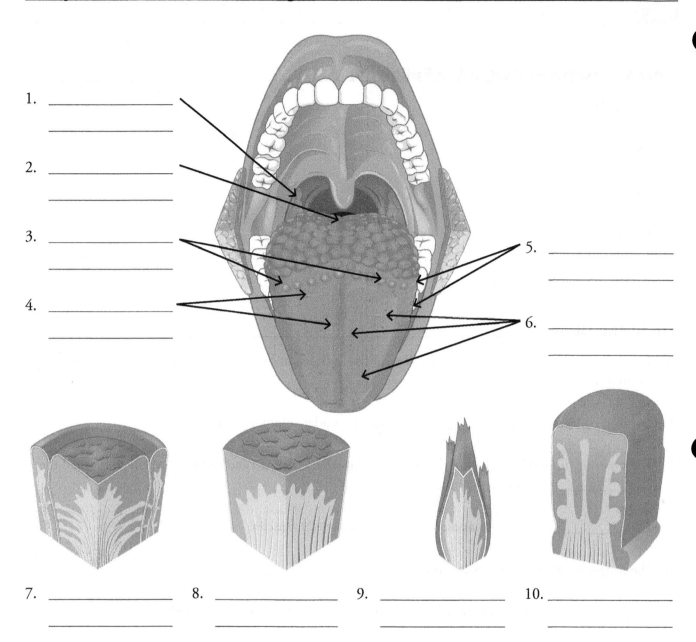

1. _____
2. _____
3. _____
4. _____
5. _____
6. _____
7. _____
8. _____
9. _____
10. _____

Digestive System & Metabolism | Pages 19–23 | Day 102–103 | Worksheet 22 | Name

Words to Know — Define the Following:

1. Periodontal ligament:

2. Cavities:

3. Saliva:

4. Tooth decay:

5. Plaque:

6. Gingivitis:

Fill in the Blank

1. In the _____ cavity is found nerves and blood vessels.

2. Both enamel and cementum contain _____, which is incorporated into their calcium-containing structures.

3. Fluoride is present in varying amounts in ordinary _____, in tea leaves, and in some foods, such as raisins and potatoes.

4. Unprotected by _____, tooth decay can become severe.

5. God designed your tooth enamel to _____ itself by incorporating minerals dissolved in your saliva.

6. The _____ produced by bacteria not only dissolve the minerals in your tooth enamel but also make it hard for teeth to recapture the lost minerals.

7. Ancient Egyptians and Babylonians — like the ones talked about in the Bible — cleaned their teeth by chewing on the frayed ends of _____.

8. The ancient Egyptians developed the oldest known recipe for toothpaste, containing dried iris flower, mint, salt, and _____.

9. There is some evidence that poor oral hygiene can lead to _____ disease.

10. Baby teeth or milk teeth are already present in a baby's _____ at birth, hidden deep beneath the gums.

Complete the Chart — Dentition: The Arrangement of the Primary Teeth

1. _____
2. _____
3. _____
4. _____
5. _____
6. _____
7. _____
8. _____
9. _____

Digestive System & Metabolism — Pages 24–27 — Day 106–107 — Worksheet 23 — Name

Words to Know — Define the Following:

1. A gland:

2. Endocrine gland:

3. Exocrine gland:

4. Parotitis:

5. Submandibular glands:

6. Saliva:

7. Amylase:

8. Xerostomia:

Fill in the Blank

1. There are _____ permanent teeth, and the buds of these teeth are present long before birth.

2. The largest of the salivary glands are the _____ glands.

3. The most common cause of parotitis is a particular viral infection called _____.

4. The _____ of our saliva start the digestive process for some of the foods we eat.

5. Saliva moistens food and this helps keep the food in a small lump, often called a "_____."

6. If not removed, plaque calcifies and hardens into _____.

7. As food is chewed, movement of the tongue, cheeks, and jaw muscles stimulates _____.

8. Inhibition of the salivary glands can occur by means of the _____ nervous system.

9. Saliva production can often be stimulated by the mere sight or smell (or even thought) of _____.

10. Chronic bad breath (_____) is associated with inadequate saliva production.

Introduction to Anatomy & Physiology 2

Complete the Chart — Salivary Glands

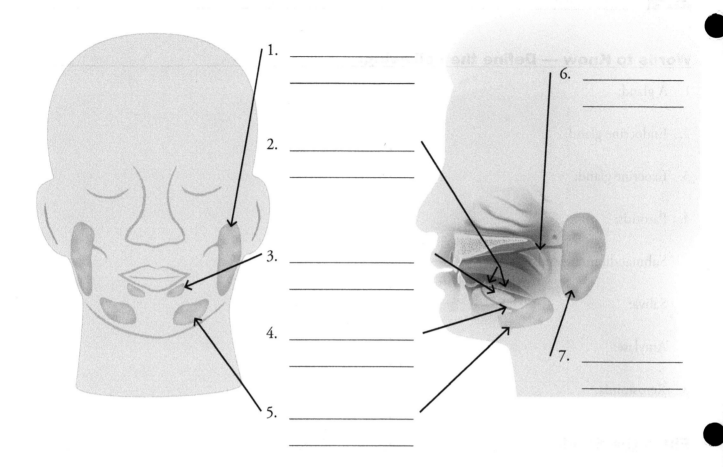

1. _____
   _____
2. _____
   _____
3. _____
   _____
4. _____
   _____
5. _____
   _____
6. _____
   _____
7. _____
   _____

Digestive System & Metabolism | Pages 28–34 | Day 110–111 | Worksheet 24 | Name

Words to Know — Define the Following:

1. Mastication:

2. Pharynx:

3. Nasopharynx:

4. Oropharynx:

5. Laryngopharynx:

6. Esophagus:

7. Sphincter:

8. Adventitia:

9. Aspiration:

10. Peristalsis:

Fill in the Blank

1. The closing of the jaw is primarily due to the actions of the powerful _____ muscle.

2. When sufficiently chewed, the bolus of food is pushed to the rear of the mouth in preparation for _____.

3. To reach the abdominal cavity, the esophagus must pass through an opening in the diaphragm known as the esophageal _____.

4. Connective tissue and blood vessels are located in the _____ along with glands that secrete mucous.

5. The upper portion of the esophagus is supplied with blood by the inferior thyroid _____.

6. Risk factors for gastroesophageal _____ disease include smoking, alcohol, diabetes, and obesity.

7. Saliva helps bind the bits of ground food into a mass called a _____.

8. When a food bolus enters the pharynx, the soft palate raises up, making a _____ between the nasal cavity and the pharynx.

9. This muscle movement along the esophagus has been described as being like a "_____."

Complete the Chart — The Esophagus

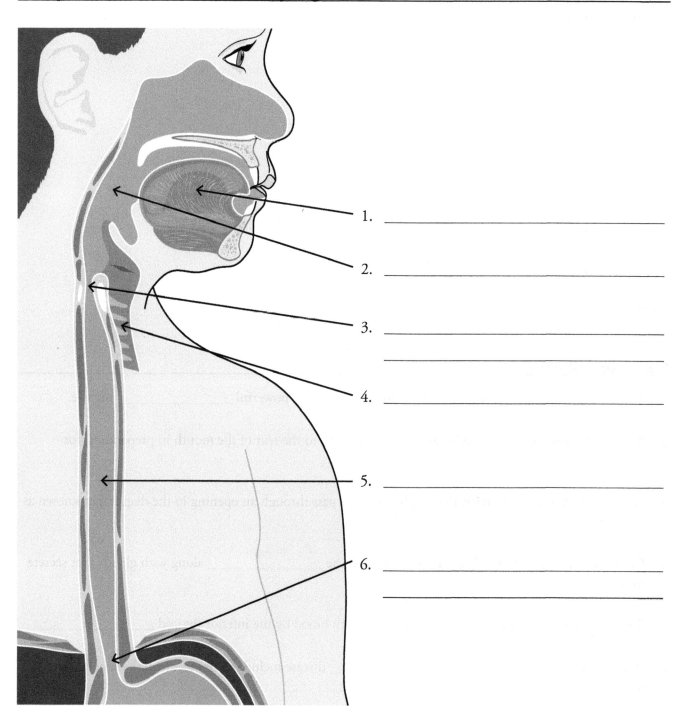

1. _____
2. _____
3. _____
   _____
4. _____
5. _____
6. _____
   _____

66 // Introduction to Anatomy & Physiology 2

Words to Know — Define the Following:

1. Greater curvature:

2. Stomach:

3. Omentum:

4. Gastric pits:

5. Parietal cell:

6. Chief cells:

7. Gastrin:

8. Intrinsic factor:

Fill in the Blank

1. Not only does the stomach secrete acid, it churns and mixes food to aid in _____.

2. When empty, the stomach lining looks wrinkled, having lots of folds, which are called _____.

3. The emptying of the stomach into the _____ intestine is controlled by the pyloric sphincter.

4. The mucosal layer of the stomach has, at its surface, a layer of special epithelial cells called _____ cells.

5. The mucus made by all those mucous cells protects the stomach lining from the corrosive effects of the very powerful _____ in the stomach.

6. A mucous layer contains a large amount of bicarbonate — a chemical found in _____ soda — which neutralizes the acid near the stomach lining.

7. Pepsinogen is _____ when produced in the chief cells and has no activity until it is secreted into the stomach.

8. If pepsinogen were an _____ enzyme, it would begin to break down proteins right away, even before being secreted into the stomach, and the proteins it would attack would be the proteins in the chief cells that produce it.

9. The chief cells also produce _____, which help break down fats in our food.

10. As food enters the stomach, the walls of the stomach are _____.

Complete the Chart — The Stomach Lining

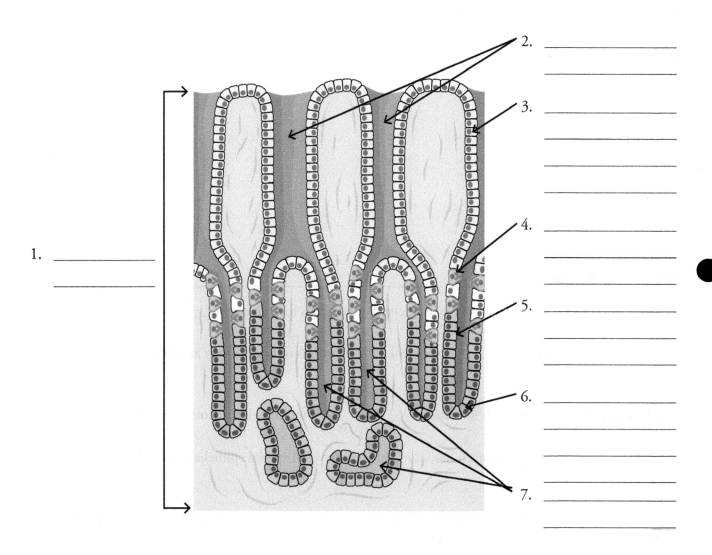

1. _____
2. _____
3. _____
4. _____
5. _____
6. _____
7. _____

| *Digestive System & Metabolism* | Pages 40–44 | Day 118–119 | Worksheet 26 | Name |

Words to Know — Define the Following:

1. Anemia:

2. Pernicious anemia:

3. Vomiting:

4. Peptic ulcer disease (PUD):

5. Treatment of PUD:

6. Pancreas:

7. Burping:

8. Gastric belch:

Fill in the Blank

1. _____ of pernicious anemia may include varying degrees of fatigue, shortness of breath with exertion, and pale skin.

2. As you might guess from the name pernicious, this sort of anemia can be very _____ if untreated.

3. When stimulated, this G cell produces a hormone called _____.

4. When _____ material is detected in the duodenum, stomach acid production is reduced.

5. Food generally remains in the stomach between two and _____ hours.

6. Vomiting results from sudden, forceful contraction of the muscles of the abdomen and the _____.

7. Emesis can result from illness, food _____, an adverse reaction to some kinds of medications, chemotherapy, radiation therapy, severe stress, motion sickness, or pregnancy, among other things.

Introduction to Anatomy & Physiology 2 // 69

8. Ongoing vomiting, besides being utterly miserable, can cause _____, which might require intravenous fluids.

9. Non-steroidal anti-inflammatory medication (NSAIDs) have the unfortunate side effect of damaging the lining of the stomach and duodenum, leading to irritation and _____.

10. Roughly 70 percent of people with PUD have stomach irritation due to H. _____.

11. The _____ activity of the stomach helps mix and grind the food as digestion continues.

12. The pancreas functions in the digestive system as an _____ gland — this is a gland that works by secreting its product into a duct.

13. The pancreas also functions as an _____ gland, a gland that secretes its products directly into the bloodstream.

14. The body of the pancreas extends laterally from the head, and it tapers into a _____.

15. We swallow _____ in small amounts as we eat, drink, and swallow the saliva in our mouths to avoid drooling.

16. When a bubble of air puts pressure on the upper part of the stomach, it triggers a reflex through the vagus nerve causing the sphincter between the esophagus and stomach to _____.

17. In a _____ belch, the air ultimately forced across the pharynx does not come up from the stomach, but is instead forcibly drawn into the upper esophagus.

Digestive System & Metabolism — Pages 45–48 — Day 122–123 — Worksheet 27 — Name

Words to Know — Define the Following:

1. Ampulla:

2. Sphincter of Oddi:

3. Pancreatitis:

4. Pancreatic islets:

5. Amylase:

6. Lipases:

7. Pancreatic cancer:

8. Liver:

Fill in the Blank

1. This ampulla is named for the 18th-century German anatomist who discovered it, Abraham _____.

2. Surgeons and gastroenterologists must know the usual anatomy very well while at the same time being alert for all the known and as yet unknown _____ that occur.

3. The majority of the pancreas consists of clusters of cells called _____.

4. When the pancreas is _____, the proenzymes leak from the secretory cells and, when activated, begin digesting proteins.

5. One of most common causes of pancreatitis is _____.

6. Other causes of pancreatitis include high levels of _____ in the blood, high levels of calcium in the blood, cystic fibrosis, infections, trauma, and certain medications.

7. The _____ in the pancreatic juice helps neutralize the acid in the stomach contents that empty into the duodenum.

8. The arterial blood supply of the pancreas comes from the _____ and superior mesenteric arteries.

9. Risk factors for pancreatic cancer include _____, diabetes, and obesity.

10. The _____ is located near the tail of the pancreas.

11. Venous drainage of the pancreas is primarily via the _____ vein.

12. In the right upper quadrant of the abdomen is found the largest organ in the digestive system, the _____.

Complete the Chart — Blood Supply to the Pancreas

1. _____
   _____
   _____

2. _____
   _____

3. _____

4. _____

5. _____
   _____
   _____

Words to Know — Define the Following:

1. Hepatocyte:

2. Bile canaliculi:

3. Portal triad:

4. Bile:

5. Jaundice:

6. Stercobilin:

7. Enterohepatic circulation:

Fill in the Blank

1. The _____ has almost 500 different functions.

2. The portal triad in the liver is not actually a triad, but a pentad, meaning it actually contains _____ things, not just three.

3. The portal vein is not really a true vein because it does not take blood to the _____.

4. Jaundice occurs in newborns when their _____ are not fully mature and is called neonatal jaundice.

5. Bile is stored in the _____ until needed.

Complete the Chart — The Pancreas, Liver, and Duodenum

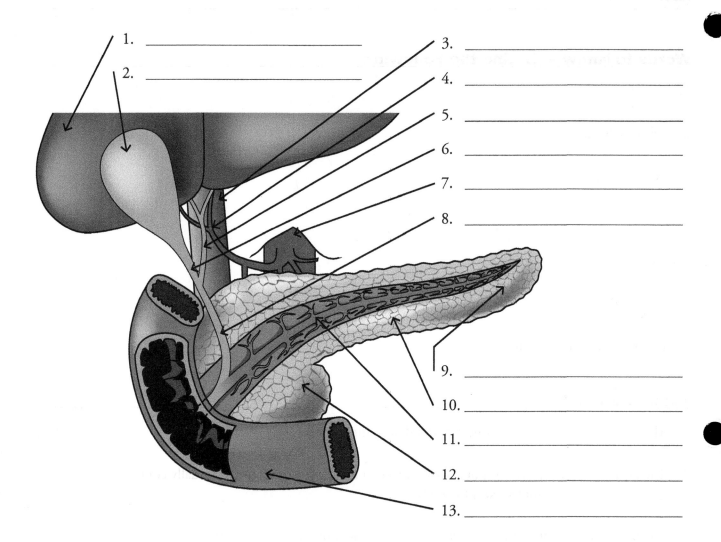

1. _____
2. _____
3. _____
4. _____
5. _____
6. _____
7. _____
8. _____
9. _____
10. _____
11. _____
12. _____
13. _____

74 // Introduction to Anatomy & Physiology 2

Digestive System & Metabolism — Pages 54–61 | Day 132–133 | Worksheet 29 | Name

Words to Know — Define the Following:

1. Metabolism:

2. Hepatitis:

3. Essential amino acids:

4. Gallstones:

5. Glycogen:

6. Small intestine:

7. Lacteal:

8. Goblet cells:

9. Enterocytes:

Fill in the Blank

1. If hepatitis persists for more than six months it is called _____ hepatitis.

2. The liver manufactures most of the proteins found in blood _____.

3. Many of the proteins involved in blood clotting, the so-called _____ factors, are manufactured in the liver.

4. A gallstone in the common bile duct can block the flow of bile and cause _____, infection, and jaundice.

5. Cholesterol is a kind of fat molecule your body's cells use to make many _____.

6. Ammonia, delivered to the liver in the portal blood, is converted by the liver into a substance called _____.

7. After passing through the pyloric sphincter, a milkshake-like mixture of partially digested food called _____ enters the small intestine.

8. Most _____ absorption takes place in the small intestine.

9. There are about 200 _____ microvilli in every square millimeter of the small intestine's surface.

10. You have inside your abdomen an absorptive surface area about the size of a tennis _____.

Complete the Chart — The Small Intestines

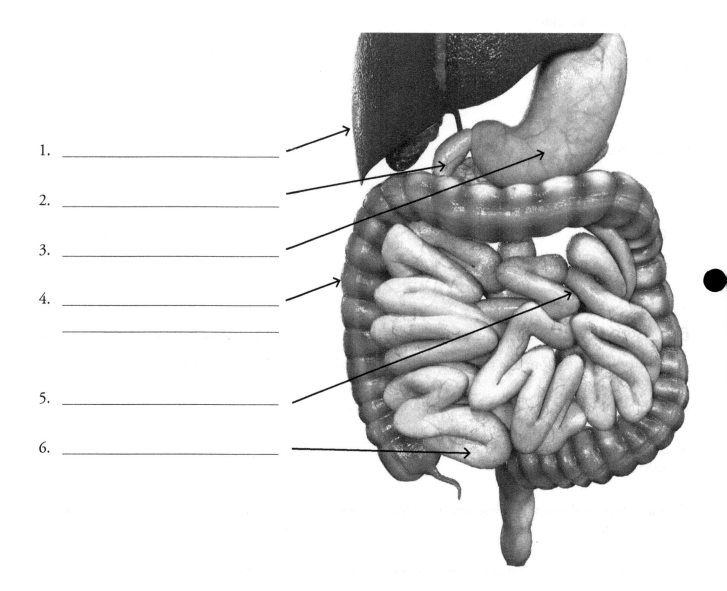

1. _____
2. _____
3. _____
4. _____
   _____
5. _____
6. _____

Digestive System & Metabolism | Page 61–65 | Day 138–139 | Worksheet 30 | Name

Words to Know — Define the Following:

1. Gastroenterologists:

2. Endoscope:

3. Endoscopy:

4. Chyme:

5. Colon:

Fill in the Blank

1. Most endoscopes even have a small channel through which the doctor can extend a pincer-like device that takes _____ — samples of tissue for microscopic examination.

2. Food enters the stomach and gets mixed with _____ acid.

3. Being made up of partially _____ food and hydrochloric acid, chyme is naturally very acidic.

4. Secretin signals the parietal cells in the stomach to lower their _____ production.

5. Enzymes brought to the duodenum from the pancreas, particularly amylase, begin to break down _____ in the chyme.

6. Amino acids are the building blocks of _____.

7. Fats, once enveloped by carrier proteins, can be safely transported through a water-based fluid.

8. Most _____ go directly to the liver without having to circulate throughout the body to reach their intended destination.

9. In some people, examination of the colon will reveal small pouches that are called _____.

10. Known as diverticulitis, this is when diverticula become _____.

Introduction to Anatomy & Physiology 2

Complete the Chart — The Large Intestine

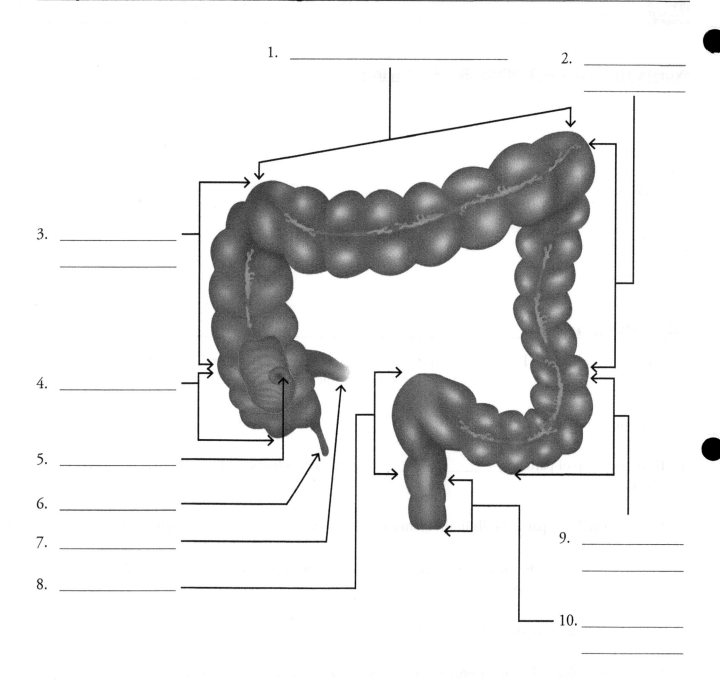

1. _____
2. _____
3. _____
4. _____
5. _____
6. _____
7. _____
8. _____
9. _____
10. _____

Digestive System & Metabolism — Pages 66–69 | Day 142–143 | Worksheet 31 | Name

Words to Know — Define the Following:

1. Taenia coli:

2. Feces:

3. Appendicitis:

4. Vestigial organ:

5. Flatulence:

6. Microbiome:

Fill in the Blank

1. Mainly just water and _____ are absorbed in the colon.

2. _____ blood is delivered to the large intestine by branches of the superior and inferior mesenteric arteries.

3. The _____ intestine removes most of the water remaining in the material coming from the small intestine.

4. The large intestine processes any remaining indigestible material into its final form called _____.

5. The appendix has a small _____, and if it cannot drain properly, bacteria get trapped in it.

6. The symptoms of appendicitis typically include _____, nausea, elevated white blood cell count, and tenderness in the right lower quadrant of the abdomen.

7. Over the last 150 years, many structures in the body were said to be leftovers from our _____ past.

8. We realize that these "vestigial" organs are not _____ but have very real functions in the body.

9. Some estimates suggest that there may be somewhere between 800 and 1,000 varieties of _____ in the GI tract.

Introduction to Anatomy & Physiology 2

10. The gas production from flatulence is the result of bacteria _____ undigested sugars and starches that enter the large intestine.

11. Between 50 and 60 percent of our feces _____ is made up of bacteria.

12. Antibiotics kill bacteria, but the antibiotics administered to kill the bacteria infecting a patient also kill lots of the "good" bacteria in the _____.

13. The presence of so many _____ bacteria can prevent more harmful bacteria from becoming a problem.

14. One of the most important products of our internal bacteria is vitamin _____.

Complete the Chart — Blood Supply to the Large Intestine

1. _____
   _____
   _____

2. _____

3. _____
   _____
   _____

4. _____
   _____

5. _____
   _____

Digestive System & Metabolism | Pages 70–74 | Day 146–147 | Worksheet 32 | Name

Words to Know — Define the Following:

1. Nutrient:

2. Macronutrients:

3. Micronutrients:

4. Essential nutrient:

5. Carbohydrates:

6. Simple carbohydrates:

7. Monosaccharide:

8. Glucose:

9. Lactose:

10. Starch:

Fill in the Blank

1. God designed the digestive system to break down and _____ all the different sorts of nutrients in our food and drink.

2. Without a proper supply of nutrients, the body could not make or _____ energy.

3. Nutrients are divided into six classes, or categories: _____, proteins, lipids, water, vitamins, and minerals.

4. Essential nutrients must be obtained from the _____.

5. Some authorities consider _____ a nutrient, while others describe only five nutrient classes.

6. The way the atoms are arranged is what defines a _____ and what gives it the properties it has.

Introduction to Anatomy & Physiology 2 // 81

7. By _____ smaller units together, much larger and more complex carbohydrates — such as starches — can be produced.

8. _____ is commonly called "fruit sugar" because it is plentiful in fruits.

9. A very familiar disaccharide is sucrose, which is ordinary table _____.

10. Complex carbohydrates are called _____.

11. Starch is made by _____; found in potatoes, rice, and wheat.

12. _____ molecules are also polymers of glucose, but they are more branched than starch.

13. When more energy is required, the glycogen in the _____ is broken down to supply glucose for the whole body.

14. Muscles deplete their glycogen during heavy exercise but are designed to rapidly rebuild it from the glucose available in the _____.

15. While not useful as a source of energy, cellulose provides much of the _____ in our diet.

Digestive System & Metabolism — Pages 75–79 — Day 150–151 — Worksheet 33 — Name

Words to Know — Define the Following:

1. Carbohydrate digestion:

2. Lactose intolerance:

3. Lactase:

4. Proteins:

5. Antibodies:

6. Insulin:

7. Amino acids:

Fill in the Blank

1. Simple _____ come from fruit, honey, and milk, as well as the sugars in candy and ice cream.

2. More _____ carbohydrates enter the body via the starches in rice, wheat, crackers, and vegetables.

3. The digestion of carbohydrates begins while you are _____.

4. The _____ in the small intestine can only absorb monosaccharides.

5. Lactose, the disaccharide sugar in milk, must be broken down into its simpler components in the small intestine in order to be _____.

6. Certain _____ populations have a higher prevalence of lactose tolerance than others.

7. Three-quarters of your dry body _____ consists of protein.

8. Proteins are certainly the most complex of the _____.

9. Amino acids are linked together by a peptide _____.

10. Generally speaking, a chain of 50 or fewer amino acids is called a _____.

Introduction to Anatomy & Physiology 2

11. A chain longer than 50 amino acids is called a _____.

12. We need protein _____ from our diet to grow, to replace worn out or damaged cells, and to make the many protein chemicals vital to cellular function and daily life.

13. The body is able to manufacture certain amino acids even if they are not consumed in adequate amounts, and these are called non-_____ amino acids.

14. Some amino acids cannot be _____ by the body and must be provided by dietary intake.

15. There are _____ essential amino acids in adults.

16. All _____ foods contain the essential amino acids we need, so they are not difficult to include in your diet.

17. If we do not obtain all the essential amino acids from our diet, the proteins requiring them cannot be made without _____ other proteins.

18. The amino acids "essential" for humans do not necessarily _____ those essential for various other living things.

Digestive System & Metabolism | Pages 79–83 | Day 154–155 | Worksheet 34 | Name

Words to Know — Define the Following:

1. Albumin:

2. Immunoglobulins:

3. Lipids:

4. Triglycerides:

5. Low-density lipoprotein:

6. Hydrophilic:

7. Hydrophobic:

Fill in the Blank

1. Our _____ has no real effect on protein, so protein digestion begins in the stomach.

2. Amino acids are the _____ blocks of proteins.

3. Amino acids not used to build proteins or make other amino acids can be turned into glucose or even lipids by the liver, but first the _____ atoms in them must be removed.

4. Every amino acid contains at least one _____ group.

5. Newborn babies are able to absorb intact _____ — those that have not been broken down — in the small intestine.

6. God designed the baby's small intestine and the mother's _____ to match each other.

7. The liver converts ammonia to a small organic molecule called _____.

8. Butter, lard, bacon fat, coconut oil, olive oil, peanut oil, and all vegetable oils are also _____.

9. Fatty acid chains with proportionately fewer hydrogen atoms are called _____ and tend to be found in oils.

10. Over the years, doctors have seen a relationship between high cholesterol and _____ disease.

Complete the Chart — Phospholipids

1. _____
   _____

2. _____
   _____

3. _____
   _____

4. _____
   _____

5. _____
   _____

6. _____

7. _____

8. _____

86 // Introduction to Anatomy & Physiology 2

Digestive System & Metabolism — Pages 84–87 — Day 158–159 — Worksheet 35 — Name

Words to Know — Define the Following:

1. Emulsification:

2. Micelles:

3. Chylomicrons:

4. Vitamins:

5. Fat-soluble vitamins:

6. Water-soluble vitamins:

7. Scurvy:

Fill in the Blank

1. There are _____ vitamins, and they are divided into two groups: fat-soluble and water-soluble.

2. In 1932, scientists discovered the vitamin C molecule, also called _____ acid, and were able to show it prevented scurvy in guinea pigs and in people.

3. Vitamin A is needed to make light-sensitive pigments in the _____.

4. You cannot overdose on the vitamin A obtained from _____ sources because they do not contain vitamin A itself but only a precursor — carotene — from which your body makes the vitamin A it needs.

5. The _____ vitamin to be discovered was vitamin A; it was creatively called "fat-soluble A."

Complete the Chart — Vitamins

VITAMIN	SOLUBILITY	DEFICIENCY DISEASE	FOOD SOURCES
Vitamin A	Fat	Night blindness, hyperkeratosis, and keratomalacia	1.
Vitamin B1 (Thiamine)	Water	Beriberi, Wernicke-Korsakoff syndrome	2.
Vitamin B2 (Riboflavin)	Water	Glossitis, angular stomatitis	3.
Vitamin B3 (Niacin)	Water	Pellagra	4.
Vitamin B5	Water	Paresthesia	5.
Vitamin B6	Water	Anemia, peripheral neuropathy	6.
Vitamin B7 (Biotin)	Water	Dermatitis, enteritis	7.
Vitamin B9 (Folate)	Water	Megaloblastic anemia, deficiency during pregnancy is associated with birth defects, such as neural tube defects	8.
Vitamin B12	Water	Pernicious anemia	9.
Vitamin C (Ascorbic acid)	Water	Scurvy	10.
Vitamin D	Fat	Rickets and osteomalacia	11.
Vitamin E	Fat	Deficiency is very rare	12.
Vitamin K	Fat	Bleeding diathesis	13.

Digestive System & Metabolism — Pages 88–91 — Day 162–164 — Worksheet 36 — Name

Words to Know — Define the Following:

1. Intrinsic factor:

2. Vitamin D:

3. Vitamin E:

4. Coagulation factors:

5. Minerals:

6. Hemoglobin molecules:

7. Iodine:

Fill in the Blank

1. Calcium is required to build and maintain strong _____.

2. Children with _____ have poorly mineralized bones that may hurt, deform as they grow, and fracture easily.

3. Rickets was a common problem until it became a popular practice to fortify _____ with vitamin D.

4. Damaged nerve cells in the _____ cannot be replaced.

5. Your best bet for nutritional health is to eat a balanced diet consisting mostly of foods that have not been _____ so much that the nutrients we need have been removed.

6. _____ and phosphorus account for the majority of the mineral content of the body.

7. Fluoride helps increase the mineral density of _____ and bones.

8. Tea contains quite a bit of _____, which is also found in raisins, carrots, potatoes, and seafood.

9. Iodine deficiency was once a big problem in many regions of the United States and Canada until the introduction of iodized _____.

Introduction to Anatomy & Physiology 2

10. Depending on conditions and activity level, many people would die without _____ after three or four days.

11. Water is lost in the process of removing _____ from the body via urine and cooling the body through the evaporation of sweat.

12. In the _____ system, water is a part of saliva, as well as a major portion of the gastric fluids and intestinal juice secreted in the GI tract.

13. It is particularly important to stay _____ before, during, and after exercise.

14. Occupational health experts recommend moderately active people in moderate conditions should drink at least a _____ of water every 15 to 20 minutes.

Complete the Chart — How Much Water is in Your Body?

1. _____
2. _____
3. _____
4. _____

| *Digestive System & Metabolism* | Pages 91–98 | Day 168–170 | Worksheet 37 | Name |

Words to Know — Define the Following:

1. Cellulose:

2. Calorie:

3. Food calories:

4. Obesity:

5. Carbohydrates:

6. Body mass index:

Fill in the Blank: Search the Chart on Page 93 for the Answers

1. Which item listed contains the highest percentage of water? _____

2. Which item listed contains the highest amount of protein? _____

3. Which item listed contains the highest amount of calories? _____

4. Which item listed contains the highest dietary fiber? _____

5. Which item listed contains the highest amount of ascorbic acid? _____

Fill in the Blank

1. Cellulose is an important part of our diet because it a major source of _____.

2. Fiber aids in the formation of _____ by producing a bulky mass on which peristalsis can act.

3. The calorie used in chemical laboratories is the amount of energy needed to raise the temperature of one gram of water one degree Celsius at one atmosphere of pressure, and a food calorie is _____ times that much energy.

4. Muscle tissue has a higher _____ rate than body fat.

5. Obesity comes down to one thing: too much _____ tissue (body fat).

Introduction to Anatomy & Physiology 2 // 91

6. Despite the bad reputation they get, we do need to consume some _____.

7. One of the best ways to "_____" your diet is to eat a variety of foods from several food groups.

8. The more you _____ the particular vegetables and fruits, grains and proteins, the more balanced your diet will be.

Complete the Chart — Nutrition Labels

Nutrition Facts
Serving Size 1 cup (228g)
Servings Per Container about 2

Amount Per Serving	
Calories 250	Calories from Fat 110
	% Daily Value*
Total Fat 12g	18%
Saturated Fat 3g	15%
Trans Fat 3g	
Cholesterol 30mg	10%
Sodium 470mg	20%
Total Carbohydrate 31g	10%
Dietary Fiber 0g	0%
Sugars 5g	
Proteins 5g	
Vitamin A	4%
Vitamin C	2%
Calcium	20%
Iron	4%

* Percent Daily Values are based on a 2,000 calorie diet. Your Daily Values may be higher or lower depending on your calorie needs:

	Calories:	2,000	2,500
Total Fat	Less than	65g	80g
Saturated Fat		25g	20g
Cholesterol	Less than	300mg	300mg
Sodium	Less than	2,400mg	2,400mg
Total Carbohydrate		300g	375g

For educational purposes only. This label does not meet the labeling requirements described in 21 CFR 101.9.

1. _____
This section is the basis for determining number of calories, amount of each nutrient, and %DVs of a food. Use it to compare a serving size to how much you actually eat. Serving sizes are given in familiar units, such as cups or pieces, followed by the metric amount, e.g., number of grams.

2. _____
If you want to manage your weight (lose, gain, or maintain), this section is especially helpful. The amount of calories is listed on the left side. The right side shows how many calories in one serving come from fat. In this example, there are 250 calories, 110 of which come from fat. The key is to balance how many calories you eat with how many calories your body uses. Tip: Remember that a product that's fat-free isn't necessarily calorie-free.

3. _____
Eating too much total fat (including saturated fat and trans fat), cholesterol, or sodium may increase your risk of certain chronic diseases, such as heart disease, some cancers, or high blood pressure. The goal is to stay below 100%DV for each of these nutrients per day.

4. _____
Americans often don't get enough dietary fiber, vitamin A, vitamin C, calcium, and iron in their diets. Eating enough of these nutrients may improve your health and help reduce the risk of some diseases and conditions.

5. _____
This section tells you whether the nutrients (total fat, sodium, dietary fiber, etc.) in one serving of food contribute a little or a lot to your total daily diet. The %DVs are based on a 2,000-calorie diet. Each listed nutrient is based on 100% of the recommended amounts for that nutrient. For example, 18% for total fat means that one serving furnishes 18% of the total amount of fat that you could eat in a day and stay within public health recommendations.

6. _____
The footnote provides information about the DVs for important nutrients, including fats, sodium, and fiber. The DVs are listed for people who eat 2,000 or 2,500 calories each day. The amounts for total fat, saturated fat, cholesterol, and sodium are maximum amounts. That means you should try to stay below the amounts listed.

Digestive System & Metabolism — Pages 99–105 — Day 173–175 — Worksheet 38 — Name

Words to Know — Define the Following:

1. Metabolism:

2. Anabolic reactions:

3. Catabolic reactions:

4. Enzymes:

5. Glycogen:

6. Cellular respiration:

7. Aerobic respiration:

8. Adenosine triphosphate:

9. Lipogenesis:

10. Triglycerides:

Fill in the Blank

1. Many substances in the body can be used as a source of fuel, but by far _____ is the most important.

2. ATP is made up of a purine molecule called _____ and a molecule of the sugar ribose.

3. The mitochondria are the _____-producing organelles — the powerhouses — of the cell.

4. The _____ cycle is a series of 8 different reactions that occur in the matrix of the mitochondria.

5. An average resting adult uses about _____% of the calories consumed simply to stay alive.

6. Without sufficient oxygen, aerobic respiration would have to stop, leaving only glycolysis — anaerobic _____ — to produce energy.

7. Any type of excess _____ can ultimately be converted into triglycerides and stored in adipose tissue throughout the body.

8. Amino acids are constantly being used to build new proteins, which is simply protein _____.

9. Protein _____ is breaking down of larger units into smaller ones.

10. It is only in situations where there is extreme calorie _____ that large amounts of protein are used for energy.

Complete the Chart — Metabolism: Where Do the Calories Go?

1. _____

2. _____

3. _____

4. _____

5. _____

Quizzes and Tests

for Use with

*Introduction to
Anatomy & Physiology 2*

The Nervous System — Quiz One — Day 24 — Quiz 1

Match the words/phrases and their definitions. (5 points each)

Autonomic nervous system   Neuroglia   Synapse
Central nervous system   Neurons   The brain
Depolarization   Sensory neurons
Motor neurons   Stimulus

1. _____ composed of the brain and the spinal cord

2. _____ excites a neuron, triggering an electrical signal called an action potential

3. _____ carry impulses away from the central nervous system

4. _____ the part of the motor division that controls the involuntary functions

5. _____ carry impulses toward the central nervous system

6. _____ the place where a neuron communicates with another neuron or with a muscle cell

7. _____ the master control center of the nervous system

8. _____ cells in nervous tissue that help protect and support the neurons

9. _____ the excitable nerve cells that transmit electrical signals

10. _____ the membrane potential becomes less and less negative, and then positive

Fill in the blank with the correct answer. (5 points each)

| experienced | nerve | neuron | replaced | signals |
| increases | nervous | peripheral | resting | Somatic |

1. The _____ is composed of three parts: the cell body, dendrites, and the axon.

2. The nervous system often compares what is sensed in the present to what has been _____ in the past.

3. Neurotransmitters are the molecules that carry the _____ across the synaptic cleft.

4. _____ means "body," so this part of the nervous system allows us to control our body's movements.

5. A _____ is made of bundles of axons located in the peripheral nervous system.

6. There exists a small electrical difference across the cell membrane, which is called the _____ membrane potential.

7. The two major divisions of the nervous system are the central nervous system (CNS) and the _____ nervous system (PNS).

8. The number of myelinated axons _____ from birth throughout childhood until adulthood.

9. Unlike most cell types in your body, neurons cannot be routinely _____.

10. The _____ tissue is the primary component of the nervous system.

The Nervous System — Quiz Two — Day 68 — Quiz 2

Match the words/phrases and their definitions. (5 points each)

Cerebellum   Primary somatosensory cortex   Synaptic plasticity

Cerebrum   Reflex   Tracts

Homeostasis   Roots

Peripheral nervous system   Spinal cord

1. _____ The ability of synapses to change their strength, resulting in the encoding of memories

2. _____ Bundles of axons in the central nervous system

3. _____ Provides a pathway for sensory information to reach the brain

4. _____ The human brain is made of four major parts; this is the largest part

5. _____ The spinal nerves connect to the spinal cord by means of two bundles of axons

6. _____ An automatic motor response triggered by a stimulus

7. _____ The body's tendency to maintain internal balance

8. _____ Helps us maintain our balance

9. _____ The portion of the nervous system outside the brain and spinal cord

10. _____ Primary is the area first in importance; somato means "body"; sensory means "input" to the brain

Introduction to Anatomy & Physiology 2

Fill in the blank with the correct answer. (5 points each)

autonomic	bridge	circulate	opposite	permanent
barrier	chemicals	nerves	optic	somatosensory

1. To prevent potentially harmful things from coming into contact with brain tissue, there is a blood brain _____.

2. Touching a piano key, sitting in a chair, drinking warm cocoa, stubbing your toe . . . all these actions produce sensory inputs that are ultimately processed in the primary _____ cortex.

3. Strokes can result in _____ neurologic damage or even death.

4. The dorsal root ganglion contains the cell bodies of sensory _____.

5. The interconnections between the four ventricles and the central canal allow the CSF to _____ around the brain and spinal cord.

6. With an _____ reflex, you remain unaware of what happened.

7. The pons (meaning "_____") is positioned between the midbrain and the medulla oblongata.

8. The _____ nerve carries nerve impulses for vision.

9. The cerebral hemispheres control the _____ sides of the body.

10. The human body was designed by the Creator God, and it is more than just a collection of _____.

The Nervous System — Quiz Three — Day 88 — Quiz 3 — Name

Match the words/phrases and their definitions. (5 points each)

Cochlea   Odorants   Tympanic membrane
Fungiform papillae   Retina   Umami
Hearing   Sound
Müller cells   Thermoreceptors

1. _____ The taste sensation produced by the amino acids, glutamate and aspartate; described as "savory"

2. _____ Respond to temperature changes

3. _____ Anterior to the vestibule; a spiral chamber made of bone

4. _____ Substances that can trigger smell

5. _____ A series of vibrations; cannot travel through a vacuum, such as in space

6. _____ Inner layer of the eye

7. _____ Our ability to convert the pressure waves (sound waves) in our environment to action

8. _____ Act like fiberoptic cables, efficiently transmitting the light that strikes the surface of the retina to the photoreceptor cells

9. _____ Marks the boundary between the external ear and the middle ear

10. _____ Mushroom-shaped; scattered over the entire surface of the tongue

Fill in the blank with the correct answer. (5 points each)

| balance | chemicals | frequency | receptor | vibrations |
| buds | eardrum | pupil | tympanic | waves |

1. If enough odorant molecules trigger the _____ cell, an impulse is sent down the entire length of the neuron.

2. Our ability to taste depends on our ability to detect, and then react to, certain _____ in our environment.

3. The vibrations transmitted from the _____ membrane are thus transferred to the oval window, and it begins to vibrate accordingly.

4. Taste _____ are found in the side walls of the foliate and circumvallate papillae.

5. The iris is a round, flat layer of smooth muscle with an opening in the middle called the _____.

6. The external auditory canal is a tube through which sound waves move toward the tympanic membrane (_____).

7. When we hear something, we are sensing sound _____ from the environment.

8. The special senses are taste, sight, smell, hearing, and _____ (or equilibrium).

9. When it reaches the tympanic membrane, a sound wave strikes the membrane and causes it to vibrate at the same _____ as the sound wave.

10. The _____ that start on the eardrum are going to be transmitted to the hammer, then to the anvil, then to the stirrup, and on to the oval window.

Digestive System & Metabolism — Quiz One — Day 135 — Quiz 1 — Name

Match the words/phrases and their definitions. (5 points each)

ampulla	digestion	omentum
amylase	esophagus	soft palate
anemia	gingivitis	
bile	metabolism	

1. _____ Found in saliva; causes the breakdown of starch in our food into sugars

2. _____ Moves during swallowing to seal off the nasal passage while food moves from the mouth into the esophagus

3. _____ A yellow-green liquid produced in the liver; made up of water, bile salts, fats, and bilirubin

4. _____ As plaque builds up, it can cause this inflammation of the gums

5. _____ Refers to all the chemical transformations that happen in our cells, both those that break down biomolecules and those that manufacture them

6. _____ Latin for "flask"; it is a sac-like enlargement of a tubular structure

7. _____ The process by which the food we take in is converted to substances needed by our bodies.

8. _____ A condition in which either the quantity or quality of a person's red blood cells is poor

9. _____ Latin word for "apron"; a double fold of peritoneal membrane

10. _____ A muscular tube that connects the pharynx to the stomach

Fill in the blank with the correct answer. (5 points each)

bolus	enamel	hormones	ingestion	parotid
dehydration	gallbladder	inactive	liver	remineralize

1. The first function of the digestive system is called _____.

2. The largest of the salivary glands are the _____ glands.

3. Pepsinogen is _____ when produced in the chief cells and has no activity until it is secreted into the stomach.

4. God designed your tooth enamel to _____ itself by incorporating minerals dissolved in your saliva.

5. Cholesterol is a kind of fat molecule your body's cells use to make many _____.

6. Bile is stored in the _____ until needed.

7. The _____ is the hardest substance in the body, and it is very durable.

8. In the right upper quadrant of the abdomen is found the largest organ in the digestive system, the _____.

9. Ongoing vomiting, besides being utterly miserable, can cause _____, which might require intravenous fluids.

10. Saliva helps bind the bits of ground food into a mass called a _____.

Digestive System & Metabolism — Quiz Two — Day 177 — Quiz 2 — Name

Match the words/phrases and their definitions. (5 points each)

albumin lipogenesis triglycerides
antibodies microbiome vitamins
calorie minerals
chyme nutrient

1. _____ Partially digested food that is ready to leave the stomach

2. _____ Assist the body in processing and utilizing other nutrients; the substance cannot be produced by the body, or at least not produced in adequate amounts

3. _____ Another class of micronutrients; inorganic substances that, like vitamins, are not used for energy

4. _____ A unit of energy

5. _____ A substance in food that is used by the body to live and grow; nutrients from food become the fuel and chemical building blocks of the body

6. _____ The storage form of lipid the body makes when we eat more calories than we burn

7. _____ Molecules produced by the immune system; these are proteins

8. _____ The process of converting nutrients into triglycerides

9. _____ The main protein found in blood plasma

10. _____ Bacteria that normally live in or on our bodies, often helping us in some way

Fill in the blank with the correct answer. (5 points each)

| absorb | chewing | fiber | metabolic | proteins |
| catabolism | digestive | intestine | proteins | retina |

1. Amino acids are the building blocks of _____.

2. The digestion of carbohydrates begins while you are _____.

3. Newborn babies are able to absorb intact _____ — those that have not been broken down — in the small intestine.

4. Muscle tissue has a higher _____ rate than body fat.

5. Vitamin A is needed to make light-sensitive pigments in the _____.

6. God designed the digestive system to break down and _____ all the different sorts of nutrients in our food and drink.

7. Protein _____ is breaking down of larger units into smaller ones.

8. Cellulose is an important part of our diet because it a major source of _____.

9. Antibiotics kill bacteria, but the antibiotics administered to kill the bacteria infecting a patient also kill lots of the "good" bacteria in the _____.

10. In the _____ system, water is a part of saliva, as well as a major portion of the gastric fluids and intestinal juice secreted in the GI tract.

The Nervous System — Test One — Day 90 — Test 1 — Name

Match the words/phrases and their definitions. (5 points each)

Autonomic nervous system   Müller cells   Spinal cord

Central nervous system   Neurons   Tracts

Cochlea   Peripheral nervous system

Homeostasis   Sound

1. _____ Act like fiberoptic cables, efficiently transmitting the light that strikes the surface of the retina to the photoreceptor cells

2. _____ The part of the motor division that controls the involuntary functions

3. _____ Provides a pathway for sensory information to reach the brain

4. _____ Anterior to the vestibule; a spiral chamber made of bone

5. _____ The body's tendency to maintain internal balance

6. _____ The excitable nerve cells that transmit electrical signals

7. _____ A series of vibrations; cannot travel through a vacuum, such as in space

8. _____ The portion of the nervous system outside the brain and spinal cord

9. _____ Bundles of axons in the central nervous system

10. _____ Composed of the brain and the spinal cord

Fill in the blank with the correct answer. (5 points each)

| autonomic | chemicals | nerve | opposite | permanent |
| barrier | increases | neuron | optic | waves |

1. When we hear something, we are sensing sound _____ from the environment.

2. To prevent potentially harmful things from coming into contact with brain tissue, there is a blood brain _____.

3. The _____ nerve carries nerve impulses for vision.

4. Our ability to taste depends on our ability to detect, and then react to, certain _____ in our environment.

5. The cerebral hemispheres control the _____ sides of the body.

6. A _____ is made of bundles of axons located in the peripheral nervous system.

7. With an _____ reflex, you remain unaware of what happened.

8. The number of myelinated axons _____ from birth throughout childhood until adulthood.

9. Strokes can result in _____ neurologic damage or even death.

10. The _____ is composed of three parts: the cell body, dendrites, and the axon.

Digestive System & Metabolism — Test Two — Day 180 — Test 2 — Name

Match the words/phrases and their definitions. (5 points each)

aspiration	mictronutrients	vitamin D
bile	peptic ulcer disease	vitamin E
digestion	saliva	
endoscope	triglycerides	

1. _____ The process by which the food we take in is converted to substances needed by our bodies.

2. _____ When food enters the airway

3. _____ A yellow-green liquid produced in the liver; made up of water, bile salts, fats, and bilirubin

4. _____ An antioxidant; able to destroy unstable chemical by-products called "free radicals" before they can oxidize and destroy cell membranes

5. _____ Called "the sunshine vitamin"; is produced by a chemical reaction in our skin when exposed to sunlight

6. _____ The storage form of lipid the body makes when we eat more calories than we burn

7. _____ Substances needed by the body, but in much smaller amounts; vitamins and minerals

8. _____ Allows the doctor to look at the lining of the GI tract; most are flexible fiber optic devices

9. _____ Occurs when there is damage to the epithelial lining of either the stomach or duodenum

10. _____ Produced by several glands in the mouth; neutralizes acids

Fill in the blank with the correct answer. (5 points each)

| acid | bicarbonate | dentin | sympathetic | useless |
| ascorbic | calories | nutrients | urea | vary |

1. _____ makes up the majority of the volume of a tooth.

2. The mucus made by all those mucous cells protects the stomach lining from the corrosive effects of the very powerful _____ in the stomach.

3. Ammonia, delivered to the liver in the portal blood, is converted by the liver into a substance called _____.

4. We realize that these "vestigial" organs are not _____, but have very real functions in the body.

5. The more you _____ the particular vegetables and fruits, grains and proteins, the more balanced your diet will be.

6. Any type of excess _____ can ultimately be converted into triglycerides and stored in adipose tissue throughout the body.

7. In 1932, scientists discovered the vitamin C molecule, also called _____ acid, and were able to show it prevented scurvy in guinea pigs and in people.

8. We need protein _____ from our diet to grow, to replace worn out or damaged cells, and to make the many protein chemicals vital to cellular function and daily life.

9. The _____ in the pancreatic juice helps neutralize the acid in the stomach contents that empty into the duodenum.

10. Inhibition of the salivary glands can occur by means of the _____ nervous system.

Answer Keys

for Use with

*Introduction to
Anatomy & Physiology 2*

The Nervous System — Worksheet Answer Keys

Worksheet 1
Words to Know: Define the Following:
1. **Sensory function:** a vast number of sensory receptors throughout the body provide input to the nervous system
2. **Motor output:** simply what the body is told to do as the result of all this information input and processing
3. **Central nervous system:** composed of the brain and the spinal cord
4. **The brain:** the master control center of the nervous system
5. **Peripheral nervous system:** portion of the nervous system outside of the central nervous system
6. **Sensory division:** carries information from the skin and muscles as well as from the major organs in the body to the central nervous system
7. **Afferent division:** another name for sensory division, and meaning "bringing toward" because it carries nerve impulses "to" or "toward" the CNS
8. **Motor division:** carries instructions from the CNS out to the body
9. **Somatic nervous system:** instructions that are carried by the motor division and taken to muscles that we can consciously control
10. **Autonomic nervous system:** the part of the motor division that controls the involuntary functions

Fill in the Blank
1. processed
2. experienced
3. Motor
4. peripheral
5. spinal
6. cranial
7. instructions
8. sensory
9. efferent
10. somatic

Complete the Chart — Automatic Nervous System
1. Stimulates saliva production
2. Slows heart beat
3. Stimulates pancreas
4. Stimulates intestinal motility
5. Decreases renin secretion (lowers blood pressure)

Worksheet 2
Words to Know: Define the Following:
1. **Neurons:** the excitable nerve cells that transmit electrical signals
2. **Stimulus:** excites a neuron, triggering an electrical signal called an action potential
3. **Neuroglia:** cells in nervous tissue that help protect and support the neurons
4. **Neurotransmitters:** the chemicals that transmit an electrical impulse from one neuron to the next
5. **Dendrites:** parts of neurons that receive inputs, and when received, an electrical signal is generated and transmitted toward the cell body
6. **Axon terminals:** where neurotransmitters are released to carry the neuron's signal on to the next cell in line
7. **Multipolar neurons:** most common type; have one axon and multiple dendrites
8. **Bipolar neurons:** have only two processes: one axon and one dendrite
9. **Unipolar neurons:** have a more unusual configuration with only one process extending from the cell body
10. **Interneurons:** means "between neurons"; carries impulses from one neuron to another within the central nervous system

Fill in the Blank
1. Epithelial
2. Connective

3. Nervous
4. Muscle
5. neuron
6. axon
7. replaced
8. efferent
9. afferent
10. glial

Complete the Chart — Neuron
1. Cell body
2. Nucleus
3. Dendrite
4. Node of Ranvier
5. Schwann cell

Worksheet 3
Words to Know: Define the Following:
1. **Myelination:** a process in which long axons are covered by a myelin sheath
2. **Schwann cells:** cells that initially indent to receive the axon, and then wrap themselves repeatedly around the axon
3. **Multiple Sclerosis (MS):** an autoimmune disease that results in the destruction of myelin sheaths in the central nervous system
4. **Neuron:** a nerve cell with dendrites and axons
5. **Motor neurons:** carry impulses away from the central nervous system
6. **Sensory neurons:** carry impulses toward the central nervous system
7. **Mixed nerves:** possess both motor and sensory fibers
8. **Nerve damage in the PNS:** does not always result in permanent loss of function
9. **Nerve damage in the CNS:** damage to the brain or spinal cord is more serious and more likely to be permanent than peripheral nerve injury
10. **Wallerian degeneration:** when distal portions of the axons begin to break down without nutrients

Fill in the Blank
1. myelin
2. nodes
3. myelination
4. increases
5. coordination
6. control
7. nerve
8. reproduce
9. creations
10. complexity

Complete the Chart — Anatomy of a Nerve
1. Epineurium
2. Axon
3. Blood vessels
4. Fascicle
5. Perineurium

Worksheet 4
Words to Know: Define the Following:
1. **Action potential:** a change in the membrane potential from negative to positive and then back again
2. **Depolarization:** the membrane potential becomes less and less negative, and then positive
3. **Threshold:** when the membrane potential reaches a certain level of depolarization to initiate the action potential
4. **All-or-none event:** when a stimulus is received, there is either a full action potential or there is no action potential at all
5. **Repolarization:** when the neuron's negative resting membrane potential is reset before another action potential can travel along that portion of the axon
6. **Continuous conduction:** when one region directly triggers the next, and the next, and the next, and so on in unmyelinated axons
7. **Saltatory conduction:** by generating local currents around the myelin sheath, the action

potential seems to "leap" from one gap to the next

8. **Graded potentials:** vary with the strength of the stimulus; the greater the stimulus, the greater number of ion channels open
9. **Synapse:** the place where a neuron communicates with another neuron or with a muscle cell
10. **Chemical synapse:** designed to transfer nerve signals by releasing special chemicals called neurotransmitters

Fill in the Blank

1. neutral
2. concentration
3. resting
4. impulse
5. depolarization
6. frequent
7. polarization
8. synapses
9. signals
10. homeostasis

Complete the Chart — Synapses Can Occur in Many Locations

1. To a dendrite
2. To the cell body
3. To another axon
4. To extracellular fluid
5. To the bloodstream

Worksheet 5

Words to Know: Define the Following:

1. **Cranial vault:** also called the cranium; the large open space in the skull
2. **Meninges:** three layers of connective tissue that cover the brain and spinal cord
3. **Periosteal:** the outermost layer of the dura attached to the inside of the cranium
4. **Meningeal:** the inner layer of the dura

5. **Pia mater:** dips down into the folds and grooves in the brain
6. **Cerebrospinal fluid (CSF):** this fluid flows around the brain and spinal cord, cushioning both
7. **Meningitis:** an inflammation of the meninges; most commonly caused by an infection
8. **Cerebrum:** the human brain is made of four major parts; this is the largest part
9. **Sulci:** the folds of the cerebrum
10. **Cerebral hemispheres:** the two halves to the cerebrum

Fill in the Blank

1. 3
2. 20
3. synapses
4. dura
5. arachnoid
6. float
7. circulate
8. ventricle
9. folds
10. lobes

Complete the Chart — Ventricles of the Brain

1. Lateral ventricles
2. Third ventricle
3. Cerebral aqueduct
4. Fourth ventricle
5. Central canal

Worksheet 6

Words to Know: Define the Following:

1. **Gray matter:** made up of the cell bodies of neurons and neuroglia; cerebral cortex
2. **White matter:** made up of both myelinated and nonmyelinated axons
3. **Corpus callosum:** a large band of white matter that connects the two cerebral hemispheres
4. **Voluntary movement:** movement you can consciously control

5. **Lateralized (lateralization):** the responsibility for certain functions rests with one hemisphere or the other
6. **Precentral gyrus:** the ridge in front of the central sulcus
7. **Pyramidal neurons:** special type of neuron with very long axons that extend all the way into the spinal cord
8. **Tracts:** bundles of axons in the central nervous system
9. **Motor homunculus:** type of visual representation of the motor regions, which means "little man"
10. **Premotor cortex:** sends appropriate signals to the primary motor cortex to get voluntary muscle movement (or movements) underway

Fill in the Blank
1. axons
2. memories
3. opposite
4. hemisphere
5. changed
6. motor
7. face
8. voluntary
9. lateralized
10. speech

Complete the Chart — Sagittal Section of the Brain
1. Corpus callosum
2. Thalamus
3. Hypothalamus
4. Medulla oblongata
5. Cerebellum

Worksheet 7
Words to Know: Define the Following:
1. **Primary somatosensory cortex:** primary is the area first in importance; somato means "body"; sensory means "input" to the brain
2. **Expressive aphasia:** the inability to speak or for people to express themselves
3. **Somatosensory association cortex:** is a processing area for sensory inputs
4. **Association areas:** the important areas of the cerebrum that receive and process information from many sources
5. **Frontal association area:** involved with learning, reasoning, planning, and abstract reasoning
6. **Visual association area:** processes visual information to allow us to understand what we are looking at
7. **Wernicke's area:** the place where language we hear is processed, allowing us to understand speech
8. **Auditory association area:** helps us distinguish between types of sounds

Fill in the Blank
1. processed
2. parietal
3. somatosensory
4. speak
5. distinguish
6. left
7. dominant

Complete the Chart — Association Areas
1. Central sulcus
2. Primary motor cortex
3. Motor association area
4. Prefrontal cortex
5. Primary auditory cortex
6. Auditory association area
7. Primary somatosensory cortex
8. Somatosensory association area
9. Visual association area
10. Primary visual cortex

Worksheet 8

Words to Know: Define the Following:

1. **Diencephalon:** the portion of the brain between the cerebrum and the brainstem; composed of the diencephalon are the thalamus and the hypothalamus
2. **Thalamus:** relays sensory input from the spinal cord to the primary somatosensory cortex; facilitates the transmission of signals from the cerebellum to the motor cortex
3. **Hypothalamus:** located below the thalamus; controls many body functions that seem fairly automatic; one of the main regulators of homeostasis
4. **Homeostasis:** the body's tendency to maintain internal balance
5. **Brain stem:** consists of the midbrain, the pons, and the medulla oblongata; provides a path for fibers extending into the spinal cord
6. **Medulla oblongata:** located below the pons, where it connects the brain to the spinal cord
7. **Reticular formation:** a net-like collection of interconnected nuclei that runs through the midbrain, pons, and medulla

Fill in the Blank

1. communication
2. sleep
3. midbrain
4. colliculi
5. bridge
6. decussate
7. stress
8. Migraine

Complete the Chart — Hypothalamus and Its Function

1. Hypothalamus
2. Pituitary gland
3. Thyrotropin-releasing hormone release, corticotropin-releasing hormone release, oxytocin release, somatostatin release
4. Circadian rhythms
5. Neuroendocrine control
6. Thermoregulation, sweating
7. Memory
8. Vasopressin release, oxytocin release
9. Increase blood pressure, pupillary dilation, shivering
10. Attention, wakefulness, memory, sleep
11. Growth hormone-releasing hormone (GHRH), feeding

Worksheet 9

Words to Know: Define the Following:

1. **Cerebellum:** helps us maintain our balance
2. **Subclavian arteries:** located under the collarbones; supply blood to the arms
3. **Subclavian steal syndrome:** blood is said to be "stolen" from the circle of Willis by the subclavian connection
4. **Circle of Willis:** named after Thomas Willis; this structure provides much protection for the brain
6. **Stroke:** occurs when cells in the brain are killed by loss of blood flow
7. **Ischemic stroke:** the most common type of stroke
8. **Aneurysm:** a dilated area on the wall of an artery; an area of weakness resembling a thin, weak section of a balloon

Fill in the Blank

1. peduncles
2. position
3. 20
4. four
5. carotid
6. uninterrupted
7. permanent
8. Blockages
9. hemorrhagic
10. brain

Complete the Chart — Blood Supply to the Brain
1. Right anterior cerebral artery
2. Anterior communicating artery
3. Right internal carotid artery
4. Right posterior communicating artery
5. Right posterior cerebral artery
6. Left middle cerebral artery
7. Basilar artery
8. Left vertebral artery

Worksheet 10
Words to Know: Define the Following:
1. **Electrocardiogram (ECG):** one way the electrical activity that makes your heart beat can be measured
2. **Electroencephalogram (EEG):** one way the electrical activity of the brain can be measured
3. **Delta waves:** this pattern is the lowest frequency and occurs in deep sleep
4. **Theta waves:** in adults, theta waves occur during meditation or drowsiness
5. **Alpha waves:** occur during wakeful but relaxed times
6. **Beta waves:** occur when the mind is active, like when concentrating or trying to communicate
7. **Sleep:** a state in which an individual achieves a degree of unconsciousness from which he or she can be aroused
8. **Rapid eye movement (REM):** the brain wave pattern during this phase of sleep is a high frequency pattern, reflecting an increase in neuronal activity
9. **Amnesia:** loss of memory; can result from trauma or severe illness
10. **Synaptic plasticity:** the ability of synapses to change their strength, resulting in the encoding of memories

Fill in the Blank
1. barrier
2. easily
3. synaptic
4. waves
5. second
6. dead
7. rapid
8. dreaming
9. memory
10. information

Complete the Chart — Brain Waves
1. Insight
   Peak focus
   Expanded consciousness
2. Alertness
   Concentration
   Cognition
3. Relaxation
   Visualization
   Creativity
4. Meditation
   Intuition
   Memory
5. Detached awareness
   Healing
   Sleep

Worksheet 11
Words to Know: Define the Following:
1. **Consciousness:** being aware of ourselves and the world around us
2. **Spinal cord:** provides a pathway for sensory information to reach the brain
3. **Spinal nerves:** 31 pairs emerging sequentially all along the length of the spinal cord
4. **Intervertebral foramina (plural):** inter- means "between," as the windows are between the vertebrae
5. **Window (foramen):** one is created where a notch on the bottom of one vertebra aligns with a notch on the top of the next one
6. **Spinal cord segment:** each area that gives rise to a spinal nerve

7. **Horns:** where the gray matter of the spinal cord projects out in several directions
8. **Ganglion:** a collection of nerve cell bodies located outside the CNS
9. **Interneurons:** neurons between neurons
10. **Lateral horns:** small horns between the dorsal and ventral horns

Fill in the Blank

1. chemicals
2. wisdom
3. spinal
4. second
5. originate
6. axons
7. posterior
8. delivered
9. sympathetic
10. columns

Complete the Chart — Spinal Nerves

1. Base of skull
2. Cervical enlargement
3. Lumbar enlargement
4. Conus medullaris
5. Cauda equina

Worksheet 12

Words to Know: Define the Following:

1. **Ascending tracts:** in the spinal cord, the sensory tracts that carry signals to the brain
2. **Descending tracts:** motor tracts that take motor output signals down the spinal cord
3. **Amyotrophic Lateral Sclerosis (ALS):** a degenerative disease of the nervous system
4. **Sporadic ALS:** in approximately 90 percent of cases where the cause of ALS is not understood
5. **Familial ALS:** inherited, with a genetic component to the illness in these situations
6. **Roots:** the spinal nerves connect to the spinal cord by means of two bundles of axons
7. **Ventral root:** contains axons of motor neurons carrying nerve signals from the CNS out to muscle and glands
8. **Dorsal root:** contains only sensory axons bringing input from sensory receptors throughout the body
9. **Corticospinal tract:** begins in the cerebral cortex and ends in the spinal cord
10. **Spinothalamic tract:** begins in the spinal cord and ends in the thalamus

Fill in the Blank

1. ascending
2. cure
3. Gehrig's
4. nerves
5. merge

Complete the Chart — Spinal Tracts

1. Sensory and Ascending (afferent) tracts
2. Dorsal white column
3. Spinocerebellar tracts
4. Spinothalamic tracts
5. Motor and Descending (efferent) tracts
6. Corticospinal tracts
7. Rubrospinal tract
8. Reticulospinal tracts
9. Vestibulospinal tract
10. Tectospinal tract

Worksheet 13

Words to Know: Define the Following:

1. **Peripheral nervous system:** the portion of the nervous system outside the brain and spinal cord
2. **Cranial nerves:** 12 pairs of nerves that emerge directly from the brain and pass through holes (called foramina) in the cranium
3. **Olfactory (I) nerves:** the first pair of cranial nerves, associated with smell
4. **Optic (II) nerves:** the second pair of cranial nerves are for sight

5. **Trigeminal neuralgia:** a pain syndrome affecting the trigeminal nerve; sometimes called tic douloureux
6. **Oculomotor nerve:** carries motor fibers to four of the six extrinsic eye muscles and the muscles of the upper eyelid
7. **Bell's palsy:** a one-sided paralysis of the muscles in the face; results from damage to the facial nerve
8. **Vagus nerve:** the only cranial nerve that extends beyond the head and neck
9. **Hypoglossal nerve:** provides almost all the motor input to the tongue, which is essential not only for speech but also swallowing
10. **Plexuses:** in areas when the spinal nerves exit they branch to form these complex networks

Fill in the Blank

1. windows
2. olfactory
3. optic
4. twelve
5. pain
6. viral
7. inside
8. output
9. lumbar
10. brachial

Complete the Chart — The Brachial Plexus

1. Musculocutaneous nerve
2. Axillary nerve
3. Radial nerve
4. Median nerve
5. Ulnar nerve

Worksheet 14

Words to Know: Define the Following:

1. **The femoral nerve:** the motor supply to muscles that flex the hip and extend the knee
2. **Carpal tunnel syndrome (CTS):** can develop when the median nerve is compressed as it runs though the wrist; presents as pain and numbness in the thumb, index, and ring fingers
3. **Shingles (herpes zoster):** a viral disease characterized by painful blisters in localized areas of the body, typically within a particular dermatome
4. **Reflex:** an automatic motor response triggered by a stimulus
5. **Somatic reflex:** results in contraction of skeletal muscles
6. **Autonomic reflex:** triggers a response in smooth muscle or glands
7. **Autonomic nervous system (ANS):** receives sensory input from sensory receptors in visceral organs — like the heart, stomach, and intestines — and blood vessels
8. **Homeostasis:** the body's ability to use many interacting mechanisms to maintain balance or "equilibrium" among its many systems
9. **Ganglion (plural, ganglia):** a collection of neuron cell bodies in the peripheral nervous system
10. **Parasympathetic nervous system:** this system is geared to support the rest and recuperation activities of the body

Fill in the Blank

1. sciatic
2. phrenic
3. dermatome
4. autonomic
5. stimuli
6. skeletal
7. sympathetic
8. preganglionic
9. plexuses
10. fight

Complete the Chart — The Fight or Flight Response

1. Threat
2. Brain
3. ACTH

4. Cortisol released
5. Adrenaline released

4. Glomerulus
5. Olfactory receptor cells

Worksheet 15
Words to Know: Define the Following:

1. **Mechanoreceptors:** sense mechanical stress, such as pressure or stretch
2. **Chemoreceptors:** respond to chemical changes, such as changes in pH (acidity)
3. **Photoreceptors:** sense light
4. **Thermoreceptors:** respond to temperature changes
5. **Olfaction:** the sense of smell
6. **Cilia:** lie on the surface of the olfactory epithelium; they are covered and protected by a thin layer of mucous
7. **Basal cells (of an epithelium):** the cells at its base, in the bottom layer
8. **Odorants:** substances that can trigger smell
9. **Triggering a sensory neuron:** in order for an olfactory sensory neuron to be triggered, an odorant molecule must reach its receptor
10. **Threshold for smell:** this is low; that means it only takes a few molecules of some odorants to trigger smell

Fill in the Blank

1. special
2. photoreceptors
3. balance
4. general
5. real
6. Proverbs
7. cilia
8. axons
9. stem
10. receptor

Complete the Chart — How Smelling Works

1. Olfactory bulb
2. Olfactory bulb neuron
3. Cribiform plate

Worksheet 16
Words to Know: Define the Following:

1. **Fungiform papillae:** mushroom-shaped; scattered over the entire surface of the tongue
2. **Gustatory epithelial cells:** means "taste"; subjected to lots of wear and tear so easily damaged with a very short life span
3. **Tastant:** stimulatory chemical that interacts with gustatory epithelial cells
4. **Afferent (sensory) fibers:** carry taste signals to the brain; mainly in two of the cranial nerves, which are the facial nerve and the glossopharyngeal nerve
5. **Bitter taste:** perceived as unpleasant; the most sensitive of the taste modalities
6. **Sweet taste:** generally pleasant; sugars are an obvious source
7. **Sour taste:** detects acidity, including citrus
8. **Salty taste:** detects inorganic salts
9. **Umami:** the taste sensation produced by the amino acids, glutamate and aspartate; described as "savory"
10. **Oleogustus:** the taste for fats

Fill in the Blank

1. chemicals
2. papillae
3. buds
4. neuron
5. receptor
6. neurotransmitter
7. tongue
8. olfactory

Complete the Chart — Taste Buds

1. Taste bud
2. Taste hairs (microvilli)
3. Taste pore
4. Gustatory cell

5. Transitional cell
6. Basal cell
7. Nerve

Worksheet 17
Words to Know: Define the Following:
1. **Auricle:** shell-shaped protrusion from the side of your head
2. **Earwax:** in the ear to keep foreign objects from reaching the delicate tympanic membrane
3. **Tympanic membrane:** marks the boundary between the external ear and the middle ear
4. **The vestibule:** the bony labyrinth's central chamber
5. **Cochlea:** anterior to the vestibule; a spiral chamber made of bone

Fill in the Blank
1. balance
2. middle
3. eardrum
4. bone
5. incus
6. vibrations
7. membranous
8. canals
9. receptor
10. vestibular

Complete the Chart — Middle Ear
1. Stabilizing ligaments
2. External acoustic meatus
3. Tympanic membrane
4. Tympanic cavity (middle ear)
5. Malleus
6. Incus
7. Stapes
8. Oval window
9. Round window
10. Auditory tube

Worksheet 18
Words to Know: Define the Following:
1. **Sound:** a series of vibrations; cannot travel through a vacuum, such as in space
2. **Pitch:** the frequency of a sound; the more waves per second, the higher the pitch
3. **Hearing:** our ability to convert the pressure waves (sound waves) in our environment to action potentials that can be transmitted to the brain
4. **The external ear:** captures sound waves (pressure waves) and directs them toward the tympanic membrane
5. **Basilar membrane:** vibrates, causing the cochlear hair cells to move against the tectorial membrane; causing movement of the hair cells generates receptor potentials
6. **Vestibular apparatus:** consists of the utricle, the saccule, and the semicircular canals; from these, signals are sent to the brain to keep it informed of the head's position in space
7. **Utricle and saccule:** positioned perpendicular to each other; therefore they are able to monitor movement in two different planes
8. **Macula:** a layer of two different types of cells: supporting cells and hair cells
9. **Vertigo:** the sensation a person experiences when he feels like he is moving but he isn't
10. **Labyrinthitis:** an inflammation of the inner ear. It may be caused by a virus

Fill in the Blank
1. waves
2. amplitude
3. Job
4. frequency
5. farther
6. bones
7. tympanic
8. tuned
9. higher
10. stroke

Complete the Chart — Auditory Path
1. Primary auditory cortex
2. Medial geniculate
3. Inferior colliculus
4. Superior olive
5. Cochlear nucleus

Worksheet 19

Words to Know: Define the Following:
1. **Corneal abrasion:** a scratch on the surface of the cornea
2. **Glaucoma:** an increase in the eye's fluid pressure that can eventually damage the retina
3. **Iris:** the colored portion of the eye; regulates the amount of light entering the eye
4. **Retina:** inner layer of the eye
5. **Cataract:** a clouding of the lens of the eye; as it progresses, there is an increasing loss of vision
6. **Müller cells:** act like fiberoptic cables, efficiently transmitting the light that strikes the surface of the retina to the photoreceptor cells
7. **Photoreceptors:** the cells that convert light into nerve impulses
8. **Color blindness:** the inability to see color or distinguish between colors
9. **Nearsighted:** one can see close objects well, but will have difficulty seeing more distant objects
10. **Farsighted:** one can see distant objects well, but closer things appear blurry and fuzzy

Fill in the Blank
1. cornea
2. pupil
3. light
4. blood
5. rods
6. cones
7. focused
8. refracted
9. 10
10. presbyopia

Complete the Chart — Retina
1. Fovea
2. Macula
3. Optic disc
4. Retinal venules
5. Retinal arterioles

Digestive System & Metabolism — Worksheet Answer Keys

Worksheet 20

Words to Know: Define the Following:

1. **Digestion:** the process by which the food we take in is converted to substances needed by our bodies
2. **Alimentary canal:** long tube that extends from the mouth to the anus; gastrointestinal tract
3. **The accessory digestive organs:** the teeth, tongue, salivary glands, liver, gallbladder, and pancreas
4. **Mechanical digestion:** the physical breaking down of food into smaller pieces
5. **Chemical digestion:** when various digestive enzymes break food down into its more basic components
6. **Absorption:** the breakdown products of chemical digestion move into the cells that line the lumen of the GI tract
7. **Elimination:** indigestible material and other substances are removed as they reach the end of the GI tract
8. **Serosa:** the outermost of the layers of the GI tract; helps provide support for the organs of the GI tract
9. **Peritoneum:** double-layered serous membrane that lines the abdominopelvic cavity; covers, at least partially, most of the organs in the abdomen
10. **Bolus:** a rounded ball of chewed food

Fill in the Blank

1. ingestion
2. Propulsion
3. feces
4. lumen
5. mucosa
6. submucosa
7. Mesenteries
8. Peritonitis
9. abdominal
10. enteric

Complete the Chart — Tissue Layers of the GI Tract

1. Lumen
2. Mucosa
3. Submucosa
4. Muscularis externa
5. Serosa

Worksheet 21

Words to Know: Define the Following:

1. **Hard palate:** bony structure covered by a mucous membrane; separates the oral cavity from the nasal cavity
2. **Soft palate:** moves during swallowing to seal off the nasal passage while food moves from the mouth into the esophagus
3. **Papillae:** the many little bumps on the surface of the tongue
4. **Tooth's neck:** the part of the tooth connecting the crown and the root
5. **Gingiva:** mucous membrane-covered connective tissue; the gums

Fill in the Blank

1. lips
2. skeletal
3. roof
4. tongue
5. hyoid
6. Taste
7. digestion
8. root
9. Enamel
10. Dentin

Complete the Chart — The Tongue

1. Palatine tonsil
2. Lingual tonsil

3. Circumvallate papillae
4. Fungiform papillae
5. Foliate papillae
6. Filiform papillae
7. Circumvallate papillae
8. Fungiform papillae
9. Foliate papillae
10. Filiform papillae

Worksheet 22
Words to Know: Define the Following:
1. **Periodontal ligament:** each tooth is secured in its socket by this complex and highly organized collection of connective tissue fibers
2. **Cavities:** holes in your tooth enamel
3. **Saliva:** produced by several glands in the mouth; neutralizes acids
4. **Tooth decay:** also known as dental caries; the result of the breaking down of the hard tissues of the tooth, primarily the enamel and the dentin
5. **Plaque:** made up of bits of food and other debris; bacteria love to live in it
6. **Gingivitis:** as plaque builds up, it can cause this inflammation of the gums

Fill in the Blank
1. pulp
2. fluoride
3. rainwater
4. enamel
5. remineralize
6. acids
7. twigs
8. pepper
9. heart
10. jaw

Complete the Chart — Dentition — The Arrangement of the Primary Teeth
1. Central incisors
2. Lateral incisors
3. Canine (cuspid)
4. First molar
5. Second molar
6. First molar
7. Canine (cuspid)
8. Lateral incisors
9. Central incisors

Worksheet 23
Words to Know: Define the Following:
1. **A gland:** an organ that produces a useful chemical substance
2. **Endocrine gland:** secretes its products directly into the bloodstream to be carried throughout the body
3. **Exocrine gland:** secretes its product by means of a duct (a small tube)
4. **Parotitis:** the inflammation of one or both of the parotid glands
5. **Submandibular glands:** empty into the mouth via the submandibular ducts; about 70 percent of saliva is produced by these glands
6. **Saliva:** a watery substance produced by the salivary glands
7. **Amylase:** found in saliva; causes the breakdown of starch in our food into sugars
8. **Xerostomia:** also known as "dry mouth syndrome"; often the direct result of an abnormally low production of saliva

Fill in the Blank
1. 32
2. parotid
3. mumps
4. enzymes
5. bolus
6. tartar
7. mechanoreceptors
8. sympathetic
9. food
10. halitosis

Complete the Chart — Salivary Glands
1. Parotid gland
2. Sublingual ducts
3. Sublingual gland
4. Submandibular duct
5. Submandibular gland
6. Parotid duct
7. Parotid gland

Worksheet 24
Words to Know: Define the Following:
1. **Mastication:** a fancy way of saying chewing; this is where mechanical digestion begins
2. **Pharynx:** a funnel-shaped tube that extends down to the level of the larynx and the esophagus; the tube that carries food and drink to your stomach
3. **Nasopharynx:** the superior portion of the larynx; extends from the rear of the nasal cavity and ends at the level of the soft palate
4. **Oropharynx:** the portion from the soft palate down to the level of the hyoid bone
5. **Laryngopharynx:** begins where the oropharynx ends, at the level of the hyoid bone; extends down to the opening of the esophagus
6. **Esophagus:** a muscular tube that connects the pharynx to the stomach
7. **Sphincter:** a ring of muscle that guards the opening at the end of a tube
8. **Adventitia:** the outer layer of the esophagus; a thin layer of connective tissue
9. **Aspiration:** when food enters the airway
10. **Peristalsis:** a series of coordinated movements of the muscles along the length of a tube, like the esophagus; the sequence of contraction and relaxation is what moves the swallowed material down the esophagus and into the stomach

Fill in the Blank
1. masseter
2. swallowing
3. hiatus
4. submucosa
5. artery
6. reflux
7. bolus
8. seal
9. wave

Complete the Chart — The Esophagus
1. Tongue
2. Pharynx
3. Upper esophageal sphincter
4. Trachea
5. Esophagus
6. Lower esophageal sphincter

Worksheet 25
Words to Know: Define the Following:
1. **Greater curvature:** the lateral aspect of the stomach
2. **Stomach:** composed of four main regions, which are the cardia, the fundus, the body, and the pyloris
3. **Omentum:** Latin word for "apron"; a double fold of peritoneal membrane
4. **Gastric pits:** an examination of the stomach lining reveals thousands of these small pits that extend down into the mucosal layer
5. **Parietal cell:** found along the walls of the gastric gland, parietal cells secrete hydrochloric acid, which begins breaking down food
6. **Chief cells:** found in the lower regions of gastric glands; produce a substance called pepsinogen
7. **Gastrin:** a hormone that aids in stimulating acid production in the stomach
8. **Intrinsic factor:** a special type of protein made by parietal cells

Fill in the Blank
1. digestion
2. rugae
3. small
4. mucous

5. acid
6. baking
7. inactive
8. active
9. lipase
10. stretched

Complete the Chart — The Stomach Lining

1. Mucous membrane
2. Gastric pits
3. Mucous cell (mucus secretion)
4. Parietal cell (HCL secretion)
5. Chief cell (Pepsin secretion)
6. Enteroendocrine cell (biogenic amines and polypeptide secretion)
7. Gastric glands

Worksheet 26

Words to Know: Define the Following:

1. **Anemia:** a condition in which either the quantity or quality of a person's red blood cells is poor
2. **Pernicious anemia:** a type of anemia associated with a deficiency of vitamin B12
3. **Vomiting:** the forceful emptying of the contents of the stomach through the mouth
4. **Peptic ulcer disease (PUD):** occurs when there is damage to the epithelial lining of either the stomach or duodenum
5. **Treatment of PUD:** includes the cessation of NSAIDs and administration of medications that lower stomach acid production
6. **Pancreas:** an accessory digestive organ that is located behind the stomach
7. **Burping:** also called "belching," or "eructation"; is the result of swallowing air
8. **Gastric belch:** a normal occurrence, ordinarily happening at least 20–25 times a day to vent the air in the stomach

Fill in the Blank

1. Symptoms

2. dangerous
3. gastrin
4. acidic
5. four
6. diaphragm
7. poisoning
8. dehydration
9. ulceration
10. pylori
11. muscle
12. exocrine
13. endocrine
14. tail
15. air
16. relax
17. supragastric

Worksheet 27

Words to Know: Define the Following:

1. **Ampulla:** Latin for "flask"; it is a sac-like enlargement of a tubular structure
2. **Sphincter of Oddi:** holds back digestive juices pooling in the ampulla of Vater until signaled by the hormone cholecystokinin to relax and open
3. **Pancreatitis:** a very serious medical condition in which the pancreas is inflamed
4. **Pancreatic islets:** or islets of Langerhans, produce the hormones that give the pancreas its endocrine functions
5. **Amylase:** enzyme that breaks down starches
6. **Lipases:** break down lipids (fats)
7. **Pancreatic cancer:** an all too common disease today; in the United States, it is presently one of the leading causes of cancer deaths
8. **Liver:** positioned just below the diaphragm on the right side of the abdomen; anteriorly, it is covered almost entirely by the rib cage

Fill in the Blank

1. Vater
2. variations

3. acini
4. inflamed
5. gallstones
6. fats
7. bicarbonate
8. celiac
9. smoking
10. spleen
11. splenic
12. liver

Complete the Chart — Blood Supply to the Pancreas

1. Common hepatic artery
2. Celiac artery
3. Aorta
4. Splenic artery
5. Superior mesenteric artery

Worksheet 28

Words to Know: Define the Following:

1. **Hepatocyte:** primary cell type found in the liver; this cell performs many different jobs, from building different types of proteins to breaking down toxins
2. **Bile canaliculi:** small ducts; into these ducts the hepatocytes secrete a substance called bile
3. **Portal triad:** located at each corner of the liver's hexagonal lobules; this triad is composed of three things: a bile duct, a small artery, and a small vein
4. **Bile:** a yellow-green liquid produced in the liver; made up of water, bile salts, fats, and bilirubin
5. **Jaundice:** a yellowish coloration of the skin due to elevated bilirubin levels
6. **Stercobilin:** the substance that gives feces its brown color
7. **Enterohepatic circulation:** the recycling of bile salts; entero- means "intestine," and hepatic means "liver"

Fill in the Blank

1. liver
2. five
3. heart
4. livers
5. gallbladder

Complete the Chart — The Relationship Between the Pancreas, Liver, and Duodenum

1. Liver
2. Gallbladder
3. Portal vein
4. Hepatic artery
5. Hepatic duct
6. Cystic duct
7. Aorta
8. Common bile duct
9. Tail of pancreas
10. Body of pancreas
11. Pancreatic duct
12. Head of pancreas
13. Duodenum

Worksheet 29

Words to Know: Define the Following:

1. **Metabolism:** refers to all the chemical transformations that happen in our cells, both those that break down biomolecules and those that manufacture them
2. **Hepatitis:** an illness characterized by inflammation of the liver; the major cause is viral infection
3. **Essential amino acids:** those amino acids the body cannot manufacture
4. **Gallstones:** small and almost pebble-like; most commonly, they are the result of having too much cholesterol or too few bile salts in the bile
5. **Glycogen:** storage form of carbohydrates which can easily be broken down into glucose when energy is needed

6. **Small intestine:** is a long tubular structure that seems to wind its way back and forth through the central portion of the abdominal cavity
7. **Lacteal:** in the center of each villus; a tiny lymphatic vessel
8. **Goblet cells:** produce the mucus that lubricates the surface of the lumen
9. **Enterocytes:** responsible for absorbing nutrients

Fill in the Blank
1. chronic
2. plasma
3. coagulation
4. inflammation
5. hormones
6. urea
7. chyme
8. nutrient
9. million
10. court

Complete the Chart — The Small Intestines
1. Liver
2. Duodenum
3. Stomach
4. Large intestine
5. Jejunum
6. Ileum

Worksheet 30
Words to Know: Define the Following:
1. **Gastroenterologists:** medical doctors that specialize in the diagnosis and treatment of diseases of the digestive system
2. **Endoscope:** allows the doctor to look at the lining of the GI tract; most are flexible fiber optic devices
3. **Endoscopy:** used to evaluate the stomach and duodenum to check for ulcers and inflammation
4. **Chyme:** partially digested food that is ready to leave the stomach

5. **Colon:** the last organ in the GI tract; the large intestine

Fill in the Blank
1. biopsies
2. hydrochloric
3. digested
4. acid
5. starches
6. proteins
7. carrier
8. nutrients
9. diverticula
10. infected

Complete the Chart — The Large Intestine
1. Transverse colon
2. Descending colon
3. Ascending colon
4. Cecum
5. Ileocecal valve
6. Appendix
7. Ileum
8. Rectum
9. Sigmoid colon
10. Anal canal

Worksheet 31
Words to Know: Define the Following:
1. **Taenia coli:** Latin for "ribbons of the colon"; these dense fibers pucker the colon into a series of large sacs called haustra
2. **Feces:** stored in the rectum until it is eliminated; stool
3. **Appendicitis:** inflammation of the appendix; can occur when the lumen of the appendix is blocked
4. **Vestigial organ:** refers to a part of the human body that has supposedly lost or changed its function during our evolutionary journey; however, our Creator did not put any useless organs in our bodies

5. **Flatulence:** gas that is expelled from the lower GI tract
6. **Microbiome:** bacteria that normally live in or on our bodies, often helping us in some way

Fill in the Blank

1. electrolytes
2. Arterial
3. large
4. feces
5. lumen
6. fever
7. evolutionary
8. useless
9. bacteria
10. fermenting
11. solids
12. intestine
13. non-pathogenic
14. K

Complete the Chart — Blood supply to the large intestine

1. Superior mesenteric artery
2. Aorta
3. Inferior mesenteric artery
4. Marginal arteries
5. Vasa recta

Worksheet 32

Words to Know: Define the Following

1. **Nutrient:** a substance in food that is used by the body to live and grow; nutrients from food become the fuel and chemical building blocks of the body
2. **Macronutrients:** the types of nutrients that the body needs in large amounts
3. **Micronutrients:** substances needed by the body, but in much smaller amounts; vitamins and minerals
4. **Essential nutrient:** something the body either cannot make on its own or cannot make fast enough to meet the body's needs
5. **Carbohydrates:** what we commonly call sugars and starches; primary use in the body is as a source of energy
6. **Simple carbohydrates:** sugars; consist of single molecular units containing only 6 or 12 carbons each, along with the appropriate number of oxygens and hydrogens
7. **Monosaccharide:** a carbohydrate consisting of a single six-carbon sugar molecule
8. **Glucose:** the favorite fuel of living cells, especially those in the brain
9. **Lactose:** the sugar found in milk
10. **Starch:** a polysaccharide made up of many glucose molecules linked together

Fill in the Blank

1. absorb
2. store
3. carbohydrates
4. diet
5. water
6. carbohydrate
7. linking
8. Fructose
9. sugar
10. polysaccharides
11. plants
12. Glycogen
13. liver
14. bloodstream
15. fiber

Worksheet 33

Words to Know: Define the Following:

1. **Carbohydrate digestion:** begins when chyme from the stomach enters the duodenum

2. **Lactose intolerance:** when people cannot fully digest and absorb the lactose in the milk and other dairy products they consume
3. **Lactase:** enzyme that breaks down lactose; is produced by cells lining the small intestine
4. **Proteins:** the building blocks of all living things; the primary structural component of all the body's tissues
5. **Antibodies:** molecules produced by the immune system; these are proteins
6. **Insulin:** controls the level of glucose in the blood; is a protein hormone
7. **Amino acids:** the basic building blocks of proteins; about 500 different amino acids exist, but there are only 20 different amino acids in the human body

Fill in the Blank

1. sugars
2. complex
3. chewing
4. enterocytes
5. absorbed
6. geographic
7. weight
8. macronutrients
9. bond
10. peptide
11. protein
12. nutrients
13. essential
14. synthesized
15. nine
16. unprocessed
17. demolishing
18. match

Worksheet 34

Words to Know: Define the Following:

1. **Albumin:** the main protein found in blood plasma

2. **Immunoglobulins:** antibodies; provide a temporarily enhanced immune system until the baby's immune system has time to make them
3. **Lipids:** include several types of compounds including fatty acids, triglycerides, phospholipids, and steroids; the term fat is often used as a synonym for lipid
4. **Triglycerides:** the storage form of lipid the body makes when we eat more calories than we burn
5. **Low-density lipoprotein:** LDL; often called "bad" cholesterol
6. **Hydrophilic:** means "water-loving"
7. **Hydrophobic:** means "water-fearing"

Fill in the Blank

1. saliva
2. building
3. nitrogen
4. amine
5. proteins
6. milk
7. urea
8. triglycerides
9. unsaturated
10. heart

Complete the Chart — Phospholipids

1. Polar head
2. Phosphate group
3. Non-polar tails
4. Polar head
5. Non-polar tails
6. Polar head
7. Non-polar tails
8. Polar head

Worksheet 35

Words to Know: Define the Following:

1. **Emulsification:** process by which the bile salts break up larger droplets into smaller and smaller droplets

2. **Micelles:** tiny collections of lipid digestive products
3. **Chylomicrons:** water-soluble droplets that are taken outside the enterocyte via exocytosis
4. **Vitamins:** assist the body in processing and utilizing other nutrients; the substance cannot be produced by the body, or at least not produced in adequate amounts
5. **Fat-soluble vitamins:** bind to lipids during digestion; can be stored by the body.
6. **Water-soluble vitamins:** are not stored in the body; excesses are simply removed from the body in the urine
7. **Scurvy:** the result of severe vitamin C deficiency; was once a leading cause of death among the crews of ships on long voyages

Fill in the Blank
1. 13
2. ascorbic
3. retina
4. plant
5. first

Complete the Chart — Vitamin Chart
1. Liver, oranges, ripe yellow fruits, leafy vegetables, carrots, pumpkin, squash, spinach, fish, soy milk, milk
2. Pork, oatmeal, brown rice, vegetables, potatoes, liver, eggs
3. Dairy products, bananas, popcorn, green beans, asparagus
4. Meat, fish, eggs, many vegetables, mushrooms, tree nuts
5. Meat, broccoli, avocados
6. Meat, vegetables, tree nuts, bananas
7. Raw egg yolk, liver, peanuts, leafy green vegetables
8. Leafy vegetables, pasta, bread, cereal, liver
9. Meat, poultry, fish, eggs, milk
10. Many fruits and vegetables, liver
11. Fish, eggs, liver, mushrooms
12. Many fruits and vegetables, nuts and seeds
13. Leafy greens, egg yolks, liver

Worksheet 36
Words to Know: Define the Following:
1. **Intrinsic factor:** produced by the parietal cells in the stomach, and is required for vitamin B12 to be absorbed in the small intestine
2. **Vitamin D:** called "the sunshine vitamin"; is produced by a chemical reaction in our skin when exposed to sunlight
3. **Vitamin E:** an antioxidant; able to destroy unstable chemical by-products called "free radicals" before they can oxidize and destroy cell membranes
4. **Coagulation factors:** proteins that make it possible for blood to clot
5. **Minerals:** another class of micronutrients; inorganic substances that, like vitamins, are not used for energy
6. **Hemoglobin molecules:** packed into red blood cells; carry oxygen all through your body
7. **Iodine:** necessary for the production of thyroid hormone; helps regulate the body's metabolic rate

Fill in the Blank
1. bones
2. rickets
3. milk
4. brain
5. processed
6. Calcium
7. teeth
8. fluoride
9. salt
10. water
11. wastes
12. digestive
13. hydrated
14. cup

Complete the Chart — How much water is in your body?
1. Adult male — 60%
2. Adult female — 55%
3. Children — 65%
4. Infant — 75%

Worksheet 37
Words to Know: Define the Following:
1. **Cellulose:** the main structural protein in plants; we take in cellulose when we eat fruits and vegetables
2. **Calorie:** a unit of energy
3. **Food calories:** used to provide the energy to keep the cells in your body functioning, not to heat water in a laboratory
4. **Obesity:** an excess of body fat; one of the most common methods for estimating this is the body mass index (BMI)
5. **Carbohydrates:** the body's primary source of energy
6. **Body mass index:** the BMI is an estimate of a person's body fat based on his or her height and weight

Fill in the Blank — Search chart on page 93:
1. Coffee, black — 99
2. Sirloin steak — 24
3. French fries — 458
4. Potato, baked — 4.8
5. Orange juice, frozen — 97

Fill in the Blank
1. fiber
2. feces
3. 1,000
4. metabolic
5. adipose
6. fat
7. balance
8. vary

Complete the Chart — Nutrition Labels
1. Serving Size
2. Amount of Calories
3. Limit These Nutrients
4. Get Enough of These Nutrients
5. Percent (%) Daily Value
6. Footnote with Daily Values (DVs)

Worksheet 38
Words to Know: Define the Following:
1. **Metabolism:** all the biochemical reactions that occur in the body to keep us alive
2. **Anabolic reactions:** those processes that take simpler molecules and build them into larger molecules or structures
3. **Catabolic reactions:** processes in which larger molecules or substances are broken down into smaller or simpler ones
4. **Enzymes:** proteins that catalyze, or speed up, a chemical reaction
5. **Glycogen:** the only storage form of carbohydrate in the human body
6. **Cellular respiration:** these are all the processes that produce ATP, the body's energy molecule
7. **Aerobic respiration:** the metabolism of glucose to produce ATP in the presence of oxygen
8. **Adenosine triphosphate:** ATP; the bonds in ATP molecules store the usable energy produced by many metabolic processes
9. **Lipogenesis:** the process of converting nutrients into triglycerides
10. **Triglycerides:** a combination of one molecule of glycerol and three fatty acid chains

Fill in the Blank
1. glucose
2. adenine
3. energy
4. Krebs
5. 70
6. respiration

7. calories
8. anabolism
9. catabolism
10. deprivation

Complete the Chart — Metabolism.

1. Brain
2. Heart
3. Liver
4. Kidney
5. Muscle

The Nervous System — Quizzes and Test Answer Keys

Quiz One
Match the words/phrases and their definitions.
1. **Central nervous system:** composed of the brain and the spinal cord
2. **Stimulus:** excites a neuron, triggering an electrical signal called an action potential
3. **Motor neurons:** carry impulses away from the central nervous system
4. **Autonomic nervous system:** the part of the motor division that controls the involuntary functions
5. **Sensory neurons:** carry impulses toward the central nervous system
6. **Synapse:** the place where a neuron communicates with another neuron or with a muscle cell
7. **The brain:** the master control center of the nervous system
8. **Neuroglia:** cells in nervous tissue that help protect and support the neurons
9. **Neurons:** the excitable nerve cells that transmit electrical signals
10. **Depolarization:** the membrane potential becomes less and less negative, and then positive

Fill in the blank with the correct answer.
1. neuron
2. experienced
3. signals
4. Somatic
5. nerve
6. resting
7. peripheral
8. increases
9. replaced
10. nervous

Quiz Two
Match the words/phrases and their definitions.
1. **Synaptic plasticity:** The ability of synapses to change their strength, resulting in the encoding of memories
2. **Tracts:** Bundles of axons in the central nervous system
3. **Spinal cord:** Provides a pathway for sensory information to reach the brain
4. **Cerebrum:** The human brain is made of four major parts; this is the largest part
5. **Roots:** The spinal nerves connect to the spinal cord by means of two bundles of axons
6. **Reflex:** An automatic motor response triggered by a stimulus
7. **Homeostasis:** The body's tendency to maintain internal balance
8. **Cerebellum:** Helps us maintain our balance
9. **Peripheral nervous system:** The portion of the nervous system outside the brain and spinal cord
10. **Primary somatosensory cortex:** Primary is the area first in importance; somato means "body"; sensory means "input" to the brain

Fill in the blank with the correct answer.
1. barrier
2. somatosensory
3. permanent
4. nerves
5. circulate
6. autonomic
7. bridge
8. optic
9. opposite
10. chemicals

Quiz Three

Match the words/phrases and their definitions.

1. **Umami:** The taste sensation produced by the amino acids, glutamate and aspartate; described as "savory"
2. **Thermoreceptors:** Respond to temperature changes
3. **Cochlea:** Anterior to the vestibule; a spiral chamber made of bone
4. **Odorants:** Substances that can trigger smell
5. **Sound:** A series of vibrations; cannot travel through a vacuum, such as in space
6. **Retina:** Inner layer of the eye
7. **Hearing:** Our ability to convert the pressure waves (sound waves) in our environment to action
8. **Müller cells:** Act like fiberoptic cables, efficiently transmitting the light that strikes the surface of the retina to the photoreceptor cells
9. **Tympanic membrane:** Marks the boundary between the external ear and the middle ear
10. **Fungiform papillae:** Mushroom-shaped; scattered over the entire surface of the tongue

Fill in the blank with the correct answer.

1. receptor
2. chemicals
3. tympanic
4. buds
5. pupil
6. eardrum
7. waves
8. balance
9. frequency
10. vibrations

Test One

Match the words/phrases and their definitions.

1. **Müller cells:** Act like fiberoptic cables, efficiently transmitting the light that strikes the surface of the retina to the photoreceptor cells
2. **Autonomic nervous system:** The part of the motor division that controls the involuntary functions
3. **Spinal cord:** Provides a pathway for sensory information to reach the brain
4. **Cochlea:** Anterior to the vestibule; a spiral chamber made of bone
5. **Homeostasis:** The body's tendency to maintain internal balance
6. **Neurons:** The excitable nerve cells that transmit electrical signals
7. **Sound:** A series of vibrations; cannot travel through a vacuum, such as in space
8. **Peripheral nervous system:** The portion of the nervous system outside the brain and spinal cord
9. **Tracts:** Bundles of axons in the central nervous system
10. **Central nervous system:** Composed of the brain and the spinal cord

Fill in the blank with the correct answer.

1. waves
2. barrier
3. optic
4. chemicals
5. opposite
6. nerve
7. autonomic
8. increases
9. permanent
10. neuron

Digestive System & Metabolism — Quizzes and Test Answer Keys

Quiz One

Match the words/phrases and their definitions.

1. **Amylase:** Found in saliva; causes the breakdown of starch in our food into sugars
2. **Soft palate:** Moves during swallowing to seal off the nasal passage while food moves from the mouth into the esophagus
3. **Bile:** A yellow-green liquid produced in the liver; made up of water, bile salts, fats, and bilirubin
4. **Gingivitis:** As plaque builds up, it can cause this inflammation of the gums
5. **Metabolism:** Refers to all the chemical transformations that happen in our cells, both those that break down biomolecules and those that manufacture them
6. **Ampulla:** Latin for "flask"; it is a sac-like enlargement of a tubular structure
7. **Digestion:** The process by which the food we take in is converted to substances needed by our bodies.
8. **Anemia:** A condition in which either the quantity or quality of a person's red blood cells is poor
9. **Omentum:** Latin word for "apron"; a double fold of peritoneal membrane
10. **Esophagus:** A muscular tube that connects the pharynx to the stomach

Fill in the blank with the correct answer.

1. ingestion
2. parotid
3. inactive
4. remineralize
5. hormones
6. gallbladder
7. enamel
8. liver
9. dehydration
10. bolus

Quiz Two

Match the words/phrases and their definitions.

1. **Chyme:** Partially digested food that is ready to leave the stomach
2. **Vitamins:** Assist the body in processing and utilizing other nutrients; the substance cannot be produced by the body, or at least not produced in adequate amounts
3. **Minerals:** Another class of micronutrients; inorganic substances that, like vitamins, are not used for energy
4. **Calorie:** A unit of energy
5. **Nutrient:** A substance in food that is used by the body to live and grow; nutrients from food become the fuel and chemical building blocks of the body
6. **Triglycerides:** The storage form of lipid the body makes when we eat more calories than we burn
7. **Antibodies:** Molecules produced by the immune system; these are proteins
8. **Lipogenesis:** The process of converting nutrients into triglycerides
9. **Albumin:** The main protein found in blood plasma
10. **Microbiome:** Bacteria that normally live in or on our bodies, often helping us in some way

Fill in the blank with the correct answer.

1. proteins
2. chewing
3. proteins
4. metabolic
5. retina
6. absorb
7. catabolism
8. fiber
9. intestine
10. digestive

Test Two

Match the words/phrases and their definitions.

1. **Digestion:** The process by which the food we take in is converted to substances needed by our bodies
2. **Aspiration:** When food enters the airway
3. **Bile:** A yellow-green liquid produced in the liver; made up of water, bile salts, fats, and bilirubin
4. **Vitamin E:** An antioxidant; able to destroy unstable chemical by-products called "free radicals" before they can oxidize and destroy cell membranes
5. **Vitamin D:** Called "the sunshine vitamin"; is produced by a chemical reaction in our skin when exposed to sunlight
6. **Triglycerides:** The storage form of lipid the body makes when we eat more calories than we burn
7. **Mictronutrients:** Substances needed by the body, but in much smaller amounts; vitamins and minerals
8. **Endoscope:** Allows the doctor to look at the lining of the GI tract; most are flexible fiber optic devices
9. **Peptic ulcer disease:** Occurs when there is damage to the epithelial lining of either the stomach or duodenum
10. **Saliva:** Produced by several glands in the mouth; neutralizes acids

Fill in the blank with the correct answer.

1. Dentin
2. acid
3. urea
4. useless
5. vary
6. calories
7. ascorbic
8. nutrients
9. bicarbonate
10. sympathetic

GENERAL SCIENCE

BASED ON OUR BEST-SELLING WONDERS OF CREATION SERIES

GENERAL SCIENCE 1
GRADE 7-12 *[1 YEAR / 1 CREDIT]*
978-1-68344-029-1

A foundational study that covers four branches of science! In this 36-week study, students will learn about oceans, astronomy, weather, and minerals. This general science curriculum serves as a great earth science option and is written from a biblical worldview perspective, emphasizing the accuracy of the Bible and the glory of God's creation all around us.

GENERAL SCIENCE 2
GRADE 7-12 *[1 YEAR / 1 CREDIT]*
978-0-89051-967-7

A unique collection touching on four fascinating sciences! Add a little variety to your science program with these full-color Wonder Series books! Learn about the landscape of caves, where you find fossils, and what processes formed the world's landscape. Discover connections with ancient civilizations and the Bible in this in depth combined program.

Daily Lesson Plans

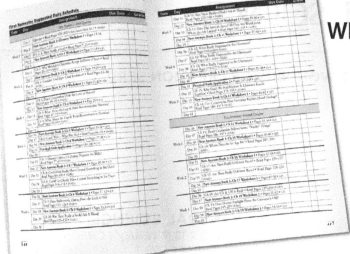

WE'VE DONE THE WORK FOR YOU!

PERFORATED & 3-HOLE PUNCHED
FLEXIBLE 180-DAY SCHEDULE
DAILY LIST OF ACTIVITIES
RECORD KEEPING

"THE TEACHER GUIDE MAKES THINGS SO MUCH EASIER AND TAKES THE GUESS WORK OUT OF IT FOR ME."

★★★★★

HOMESCHOOL

Master Books® Homeschool Curriculum

Faith-Building Books & Resources
Parent-Friendly Lesson Plans
Biblically-Based Worldview
Affordably Priced

Master Books® is the leading publisher of books and resources based upon a Biblical worldview that points to God as our Creator. Now the books you love, from the authors you trust like Ken Ham, Michael Farris, Tommy Mitchell, and many more are available as a homeschool curriculum.

MASTERBOOKS.COM
— *Where Faith Grows!* —